TAIPEI 101

TAIPEI 101

Georges Binder

images
Publishing

Published in Australia in 2008 by
The Images Publishing Group Pty Ltd
ABN 89 059 734 431
6 Bastow Place, Mulgrave, Victoria 3170, Australia
Tel: +61 3 9561 5544 Fax: +61 3 9561 4860
books@imagespublishing.com
www.imagespublishing.com

The Images Publishing Group Reference Number: 797

National Library of Australia Cataloguing-in-Publication entry:

Title:	Taipei 101 : editor Georges Binder.
ISBN:	9781864702484 (hbk.)
Subjects:	Tall buildings-Taiwan-Taipei-Designs and plans.
	Tall buildings-Taiwan-Taipei-Pictorial works.
	Taipei (Taiwan)-Buildings, structures, etc.
	Taipei (Taiwan)-Pictorial works.
Other Authors/Contributors:	Binder, Georges.
Dewey Number:	720.4830951249

Edited by Robyn Beaver

Designed by The Graphic Image Studio Pty Ltd, Mulgrave, Australia
www.tgis.com.au

Pre-publishing services by Splitting Image Colour Studio Pty Ltd, Australia
Printed by Everbest Printing Co. Ltd. in Hong Kong/China

IMAGES has included on its website a page for special notices in relation to this and our other publications. Please visit www.imagespublishing.com.

Contents

TAIPEI 101

TAIPEI 101 is more than a remarkable technological achievement; it symbolises the growth and economic prosperity of Taiwan and stands as an invitation to the rest of the world to do business in, and with, Taiwan. From the outset, the aims of the architects, developers and stakeholders were clear; they are identified in the building name itself:

Technology, **A**rt, **I**nnovation, **P**eople, **E**nvironment and **I**dentity.

Technology: TAIPEI 101 speaks for itself while its users benefit from the latest technology in media and communications.

Art: TAIPEI 101 features pieces from international artists while the building is treated as a giant sculpture.

Innovation: TAIPEI 101 constantly displays its innovative ways to the people of Taipei, from its ever-changing light displays, to the promotion of new merchandise and special publicity campaigns.

People: TAIPEI 101 provides outstanding work and leisure areas that attract residents and visitors alike.

Environment: TAIPEI 101 is a meticulously planned urban insertion. Transparent glass walls surround the mixed-use retail podium that is alive with activity day and night; in the basement a well planned drive-through and taxi waiting areas have been implemented.

Identity: TAIPEI 101's iconic nature is the result of developer Harace Lin and architect C.Y. Lee's desire to not only give the project a strong Oriental identity, using the best technology available from all over the world, but to create a building that specifically could only be designed for their own place with references to their own culture. There is no doubt that their ambition has been achieved in this miracle of modern technology.

The world's tallest buildings:

A HISTORICAL OVERVIEW

For more than a century, the world's tallest building was a North American office building located in either New York or in Chicago. In 1998, the location shifted to Asia, with the completion of the Petronas Towers in Kuala Lumpur, Malaysia. In 2004, TAIPEI 101, in Taipei, Taiwan became the world's tallest building.

The world's tallest buildings, 1890-2004

1890	World Building, New York, New York, USA, 94 metres (309 feet), demolished
1892	Masonic Temple, Chicago, Illinois, USA, 92 metres (302 feet), demolished
1894	Manhattan Life Insurance Company Building, New York, New York, USA, 106 metres (348 feet), demolished
1898	St. Paul Building, New York, New York, USA, 96 metres (315 feet), demolished
1899	Park Row Building, New York, New York, USA, 119 metres (391 feet)
1898	Singer Building, New York, New York, USA, 187 metres (612 feet), demolished
1909	Metropolitan Life Insurance Tower, New York, New York, USA, 213 metres (700 feet)
1913	Woolworth Building, New York, New York, USA, 241 metres (792 feet)
1930	Bank of Manhattan Company Building, New York, New York, USA, 283 metres (927 feet)
1930	Chrysler Building, New York, New York, USA, 319 metres (1046 feet)
1931	Empire State Building, New York, New York, USA, 381 metres (1250 feet)
1971/73	World Trade Center, New York, New York, USA, 417/415 metres (1368/1362 feet), demolished
1974	Sears Tower, Chicago, Illinois, USA, 442 metres (1451 feet)
1998	Petronas Towers, Kuala Lumpur, Malaysia, 452 metres (1483 feet)
2004	TAIPEI 101, Taipei, Taiwan, 508 metres (1667 feet)

Tall building rivalry · In 1930-32, the race for height among New York skyscrapers heated up between downtown's Bank of Manhattan Company Building and Cities Service Building, and midtown's Chrysler Building and Empire State Building. So intense was the rivalry that the spire of the Chrysler Building was secretly assembled within the building crown, and raised into place just as the tower was finished, keeping the final height a secret until the last minute and ensuring that the 40 Wall Street project did not hold the world's tallest building title for long.

Beginning with the midtown-versus-downtown New York rivalry in the 1930s, the race for the world's tallest building has continued. It later involved different cities within the same country - the World Trade Center in New York (1973) was surpassed a year later by the Sears Tower in Chicago by less than 30 metres (100 feet). It took another quarter of a century for the race to continue, this time involving different continents competing against each other. Asia has emerged the winner at this time, but the trend of secrecy in relation to the height of the spire continues with super-tall projects currently under progress in the Middle East.

The 1930 77-storey 319-metre Chrysler Building in New York, designed by William Van Alen and once the world's tallest building, has become the quintessential North American skyscraper, proving that good design remains when world records do not.

Super-tall buildings in Asia · In 1989, the 367.4-metre 70-storey Bank of China Tower in Hong Kong not only became the tallest building ever built outside North America, but it was the first such building to be designed by an architect of Asian origin – I.M. Pei. More tall buildings in Asia quickly followed: in 1992 the 78-storey, 374-metre Central Plaza, also in Hong Kong, designed by Dennis Lau of Ng Chun Man Architects and Engineers (HK) Ltd., the first Asian-based architect to complete a super-tall building. In 1997 the 73-storey, 346-metre The Center in Hong Kong and the 80-storey, 390.2-metre CITIC Plaza (originally Sky Central Plaza) in Guangzhou, China, were also designed by Dennis Lau & Ng Chun Man Architects & Engineers (HK) Ltd.

The 1992 78-storey Central Plaza in Hong Kong (374 metres) was the first super-tall building designed by an Asian-based architect, Dennis Lau of Ng Chun Man Architects and Engineers (HK) Ltd.

It has been more than a decade since Asia became the world leader in tall buildings, after a century of North American supremacy. Asia's dominance began in 1998 with the completion of the 88-storey Petronas Twin Towers (originally the Petronas Towers), designed by Cesar Pelli & Associates. The Petronas Towers established a new world record height of 452 metres and officially surpassed the 110-storey, 442-metre Sears Tower in Chicago (designed by Bruce Graham of Skidmore, Owings & Merrill), which had held the height record since 1974. A year after the completion of the Petronas Twin Towers, the 88-storey, 421-metre Jin Mao Tower (originally the Jin Mao Building), designed by Adrian Smith of Skidmore, Owings & Merrill, became China's tallest building.

At the turn of the new millennium, in 1999 and 2000, the Middle East began to make its mark. It was one of the few regions of the world where super-tall buildings were being built. In 1999 the 321-metre, 52-storey Burj Al Arab hotel (designed by WS Atkins) was completed; the twin Jumeirah Emirates Towers (originally Emirates Towers), designed by NORR Group Consultants International (Hazel Wong) with Hyder Consulting Middle East Ltd. was completed in 2000. The 355-metre tower office tower and the 309-metre lower hotel tower are built above a shopping mall. Today, Dubai is home to a series of super-tall towers under construction, predominantly residential or mixed-use, including the Burj Dubai designed by Adrian Smith. The mixed-use Burj Dubai, scheduled for completion in 2009 should become the new world's tallest building by a large margin while several residential towers, from 300 metres to more than 500 metres are already under progress at the Dubai Marina. At the time of writing, about 15 high-rise buildings of more than 300 metres are under construction in Dubai alone; in 1995 there were fewer than 15 such buildings in the entire world.

A recent view of the Pudong area of Shanghai, China showing the 101-storey Shanghai World Financial Center designed by Kohn Pedersen Fox Associates (left) and the 88-storey Jin Mao Tower designed by Adrian Smith of Skidmore, Owings & Merrill (right).

In the post-9/11 era, Hong Kong was again on the tall building map with the completion in 2003 of the 412-metre, 88-storey Two International Finance Centre - known as 'Two ifc' - designed by Cesar Pelli & Associates and built on reclaimed land, as is the case for many of the super-tall buildings in the city. Today, in Kowloon and also built on reclaimed land, the mixed-use, 118-storey, 490-metre International Commerce Centre, designed by Kohn Pedersen Fox Associates, will become one of the few buildings in the world with more than 100 storeys. A Ritz-Carlton hotel will be located in the upper levels of the project. The long-awaited Shanghai World Financial Center, at 101 storeys and 492 metres, graces the Shanghai skyline. The mixed-use project, designed by Kohn Pedersen Fox Associates, and developed by Mori Building of Tokyo, comprises a shopping mall at its base and a Park Hyatt hotel atop the building, making this hotel, when completed, the world's tallest hospitality project. Other locations in mainland China where super-tall buildings are currently being constructed include, among others, Beijing, Guangzhou and Nanjing.

Super-tall buildings in Europe · In the West, few tall buildings have appeared since Frankfurt's 299-metre, 51-storey Commerzbank Tower, designed by Foster and Partners, was completed in 1997. A series of projects are nearing completion in Moscow, Russia where the mixed-use Federation Tower should become Europe's tallest building. In Madrid, Spain four 236- to 250-metre-high office towers are being completed at the outskirts of the city. In London, England super-tall projects under progress include the 288-metre, 64-storey Pinnacle, designed by Kohn Pedersen Fox Associates and the 306-metre, 66-storey mixed-use Shard London Bridge designed by Renzo Piano Building Workshop. London is one of the few European cities to have implemented an urban policy that encourages tall and super-tall buildings on a large scale with the goal of adding density to certain areas in the city centre. Paris, France may follow in the years to come, although at a slower pace, with the historical high-rise district La Défense also scheduled to accommodate several 300- to 400-metre-high office towers.

The 95-storey Federation Tower in Moscow, Russia is scheduled to become Europe's tallest building.
Architect: Prof. Peter Schweger von ASP Schweger Assoziierte Gesamtplanung, Hamburg and Sergei Tchoban von nps tchoban voss, Berlin

Super-tall buildings in the USA · The United States, once a major force in super-tall projects, lags behind at this point, despite the 415-metre Trump International Hotel & Tower in Chicago, Illinois designed by Adrian Smith. The 92-storey hotel and residential tower, adjacent to the Chicago River, is scheduled for completion in 2009. It is the first super-tall building in North America built without a single square metre of office space. Other super-tall projects under progress in the United States include the Freedom Tower in New York, designed by David Childs, consulting design partner of Skidmore, Owings & Merrill and due for completion in 2013. It will be the first tower to be completed on the site of the former World Trade Center and was originally expected to become the tallest building in North America at 541 metres.

The future WTC in New York City, due for completion around 2013. Projects are designed by Skidmore, Owings & Merrill, Foster + Associates, Rogers Stirk Harbour + Partners and Maki & Associates.

T&C Tower, Kaohsiung, Taiwan.

It will also be one of the few Western projects to incorporate the type of numerical symbology that is often found in Asian buildings: the height will reach 1776 feet - 1776 representing the foundation year of the United States of America. Eventually, the tallest North American building may well be the Chicago Spire. Designed by Santiago Calatrava, this residential tower in Chicago could reach 609 metres.

Taiwan and super-tall buildings · Where does Taiwan stand in terms of super-tall structures? As a country with few tall buildings until the early 1990s, Taiwan made its first foray into tall building circles when the 222-metre Grand 50 Tower in Kaohsiung and the 244-metre Shin Kong Life Tower in Taipei became Taiwan's tallest buildings when completed in 1992 and 1993 respectively. In 1998 the 85-storey mixed-use T&C Tower, also in Kaohsiung and designed by C.Y. Lee & Partners, became the country's tallest building at 348 metres, placing Taiwan among the few countries to erect super-tall buildings at that time.

TAIPEI 101
World's Tallest Building
508 meters, 1,667 feet
According to criteria established by the Council on Tall Buildings and Urban Habitat
October 8, 2004
Ron Klemencic
Chairman, CTBUH
David Maola
Executive Director, CTBUH
Council on Tall Buildings and Urban Habitat
Taipei Financial Center Corporation
Developer
C. Y. Lee and Partners
Architect

Development history of
TAIPEI 101

Team builder/dream builder · A graduate of Tamkang University and Vice Chairman of the Hung Kuo Group, Harace Hong-Min Lin is President of TAIPEI 101. While most young boys were playing with toy cars, Harace Lin was probably playing with construction blocks, manifesting his early interest in tall structures. A fitting start for the man behind the idea of creating the tallest structure ever built. His father, who was involved in the real estate business, instilled in him a passion for building. 'Not especially building tall, but building something that would last, something that you could pass on', says Harace Lin. Before TAIPEI 101, Harace Lin and his brother Eric had an early dream of building a 126-storey tower - also designed by C.Y. Lee & Partners and Thornton Tomasetti - on one of the most sought-after sites in Taipei, the location of the SOGO department store along Chung Xiao East Road. That tower, designed in the form of a blossoming flower, did not pass the environmental impact analysis when proposed in 1989.

Harace Lin, President, TAIPEI 101.

126-storey early super-tall project for Taipei by Harace and Eric Lin.

Team builder · Harace Lin believes in good timing and coincidence in creating imaginative, major real estate achievements. In 1997, when the property market was prosperous and capital costs were low, the Mayor of Taipei invited developers, including Harace Lin, to bid for a major site. The newly vacated area, Xin Yi Planned District, was designed to become a major multifunctional area in the Taiwanese capital. In order to win this 70-year BOT (Build Operate Transfer) project, Harace Lin first assembled a team comprising major Taiwanese financial institutions, headed by China Development Industrial Bank and including Cathay Insurance, Shin Kong Insurance and China Insurance. Relationships matter to Harace Lin and his ties with the leaders of these financial institutions are far more than business ties - all these financial tycoons have been golf mates of his for many years. But experience, trust and confidence were not enough when they realised that, according to Taiwanese regulations, insurance companies are prohibited from investing more than 5 percent of their assets in a single project. To overcome this, Harace Lin immediately put together a wider financial team, which was strong enough to bid for a major piece of land and would meet all the regulations. Ten companies were approached - later eleven, with the Taiwan Stock Exchange becoming a 5 percent shareholder, a requirement of the Taipei City Government's tender document. The result was that TAIPEI 101 became the first joint effort between the Taipei City Government and the private sector. Eventually, backed by companies forming a strong team put in place by Harace Lin, the Taipei Financial Center Corporation was established in July 1997 and has been the land developer responsible for the planning, implementation, operation and management of TAIPEI 101. The current investors include China Development Industrial Bank, China United Trust & Investment Corporation, Chinatrust Financial Holding, Cathay Financial Holding, Chunghwa Telecom, Mega Holdings, Shin Kong Insurance, Taiwan Stock Exchange, Walsin Lihwa, First Financial Holding, Hua Nan Financial Holding and Taishin Bank.

Originally, seven developers - or teams of developers - won the right to bid in 1997 for the tender. Harace Lin's team emerged as the winner, with the highest amount proposed at NT$20,688,890,000 and the project design reviewed by a committee put in place by the Taipei City Government.

Dream builder, or how the project turned into the world's tallest · The initial project did not call for the tallest building in the world; it required that a 59-storey project be developed. To accommodate the needs of the early major building occupants - China Life Insurance and the Taiwan Stock Exchange - the first design suggested was a 66-storey, 273-metre project, flanked by twin 20-storey towers. The 66-20-20 design proposal was rejected by some future occupants - it was described by some as 'two small dogs standing guard next to their master'. In response, Harace Lin proposed stacking each lower building on top of the 66-storey tower, but soon discovered that this solution could potentially result in the erection of the world's tallest structure. But how tall and exactly how many storeys would it be? '100 would be a perfect number', explains Harace Lin. 'But I wanted it to be more than perfect', he adds proudly with a smile. After much thought, he arrived at the 101 figure that would later become part of the TAIPEI 101 brand name, symbolising both 101 levels and today's digital world. In fact, between the 66-storey project and the final 101-storey design version, several other projects were considered, including an 88-storey, 428-metre tower and a 100-storey, 488-metre tower.

The TAIPEI 101 project has raised the overall quality of real estate in Taiwan, says Harace Lin. From state-of-the-art power supply and communication systems to raised floors and high floor-to-ceiling heights, TAIPEI 101 features the best of both Western and Eastern grade-A office buildings. Attention to interior finishes, the inclusion of artworks of international artists throughout the project and the quality landscaping that surrounds the project are all world-class. TAIPEI 101 was designed to accommodate tenants with the highest standards and expectations - among its major occupants are companies including ABN AMRO, ING Insurance, KPMG and the Taiwan Stock Exchange.

The initial 66-storey 273-metre project proposed at the time the bidding for the TAIPEI 101 land was won.
Architect: C.Y. Lee & Partners Architects/Planners

A mixed-use ensemble · The urban authorities did not authorise a hotel or residential units to be part of the project. Although the regulations were changed at a later stage to allow the inclusion of a hotel, it was too late to include a hospitality project within the 101-storey tower. However, TAIPEI 101 is a true mixed-use ensemble, encompassing the largest shopping mall (77,000 square metres) in Taipei, with famous brand names including Cartier, Dior, Prada, Tiffany & Co and Louis Vuitton, to name just a few. When asked to explain why vertical malls work well in Asia and not so well in other parts of the world, Harace Lin takes his pencil and demonstrates. He shows how people would enter the mall both from the bottom, in a most logical and usual way, and from 'the air' with people coming out of the 101-storey tower and exiting at levels four and five. People move efficiently within the huge retail and entertainment venue, of which level four, with its landscaped square, is a major component of the urban life created within the TAIPEI 101 project. This fourth level is 'a destination level', explains Harace Lin, with its numerous restaurants, bookshop and a variety of corner boutiques. It blends with the indoor plaza level, giving it the flavour of an outdoor place with its benches and landscaped alleys.

When asked about the idea of creating a mall evolving along a large curved atrium - an early mall design was strictly linear - Harace Lin again uses his pencil in order to explain that people walking through a curved mall would have more opportunities to see

more places and 'the more they see, the more they buy', he explains. Of course, the mixed-use project is complemented by an observatory on the 89th level, a Taiwan specialty gallery on the 88th level, an outdoor terrace on the 91st level, restaurants on levels 85 and 86 and the conference centre on level 36.

Creating the best business precinct and the most imaginative retail environment to accommodate the needs of financial tenants and high-end international brands was obviously Harace Lin's professional ambition, but creating a multifunctional environment with a sense of place is probably even closer to what he had in mind with his early wish to build 'something that you could pass on'.

Creating an East-West dialogue · According to Harace Lin, a building conveys two messages: culture and technology. Every building reflects local culture and tall buildings require technology brought in from a variety of foreign places. Harace Lin was familiar with the 85-storey T&C Tower in Kaohsiung (Taiwan), designed by C.Y Lee & Partners, so he knew that by working with architect C.Y. Lee, he could convey the message he had in mind. Since the architect had already successfully completed an 85-storey building, he was already familiar with the technology required to build such a project. Significantly, Harace Lin and C.Y. Lee had teamed up several times during the last two decades to create real estate and architecture, fulfilling both market requirements and their respective wishes to apply their Chinese cultural roots within the Taiwanese contemporary community. Their collaborative projects included the 19-storey headquarters building built in 1989 for Hung Kuo, the real estate development company of which Harace Lin is Vice-Chairman.

In TAIPEI 101, Harace Lin wished to create a dialogue between Western and Oriental culture, which clearly explains his choice in engaging architect C.Y. Lee. Harace Lin also insisted that TAIPEI 101 should feature the best available Western technology. The mega-structure structural systems, the evacuation systems with the refuge floors (which were a first for Taiwan and a safety measure first found in Hong Kong and mainland China), the double deck elevators and the sky-lobbies, the pressurised elevators for optimal human comfort (also a world first), access to the observation deck and the tuned mass damper atop the building are just a few of these impressive technical achievements.

The coin test · Harace Lin obviously likes to meet with people and create places for people to meet, like the covered City Square on the fourth level of the mall. He is also a serious man: he is a developer and he has built the world's tallest building, a 508-metre-tall structure comprising 357,719 square metres of gross usable space. A serious man he may be, but the gleam in his eye when performing the 'coin test' in the elevator shows another dimension: descending from the 101st level, the top-most level of TAIPEI 101 used for private functions by its President, Harace Lin likes to balance a coin on its edge on the floor of the pressurised observatory cabin while descending in the world's fastest elevator. The joy evident on his face when arriving at the base of the tower and seeing - like everybody else in the elevator cab - that the coin is still in place on its edge is a wonderful experience. Efficiency and smoothness, like the TAIPEI 101 elevators, are words that come to mind after a meeting with Harace Lin. Efficiency and smoothness define Harace Lin and that was his desire for TAIPEI 101.

The architecture of

TAIPEI 101

Taipei-based architect C.Y. Lee was born in 1938 in Guangdong, China and arrived in Taiwan with his parents in 1949. He received a Bachelor of Science in Architectural Engineering in 1961 from National Cheng Kung University (Tainan, Taiwan) and his Master of Art in Architecture degree from Princeton University (New Jersey, USA) in 1966. After working with a series of American architects, including William L. Pereira from 1971 to 1977, C.Y. Lee returned to Taiwan in 1978. His first major commission in Taiwan was the 138,000-square-metre, 15-storey Asia World Plaza in Taipei, a large curved concrete structure organised around a vast atrium. This substantial mixed-use project, comprising a department store and an 800-room hotel, established C.Y. Lee as an architect able to successfully handle large-scale projects. While this first Taiwanese project, with its glass-enclosed elevators, owed much to his American experience, it was also his first opportunity to integrate aspects of his own cultural heritage into a building.

C.Y. Lee, Architect, C.Y. Lee & Partners Architects/Planners.

C.Y. Lee's partner C.P. Wang was born in 1947 in Beijing, China and, together with his family, arrived in Taiwan in 1949. He earned a Bachelor of Science in Architectural Engineering from the Taichung-based Tunghai University. Wang then continued his studies in the USA where he graduated from Washington University in 1973.

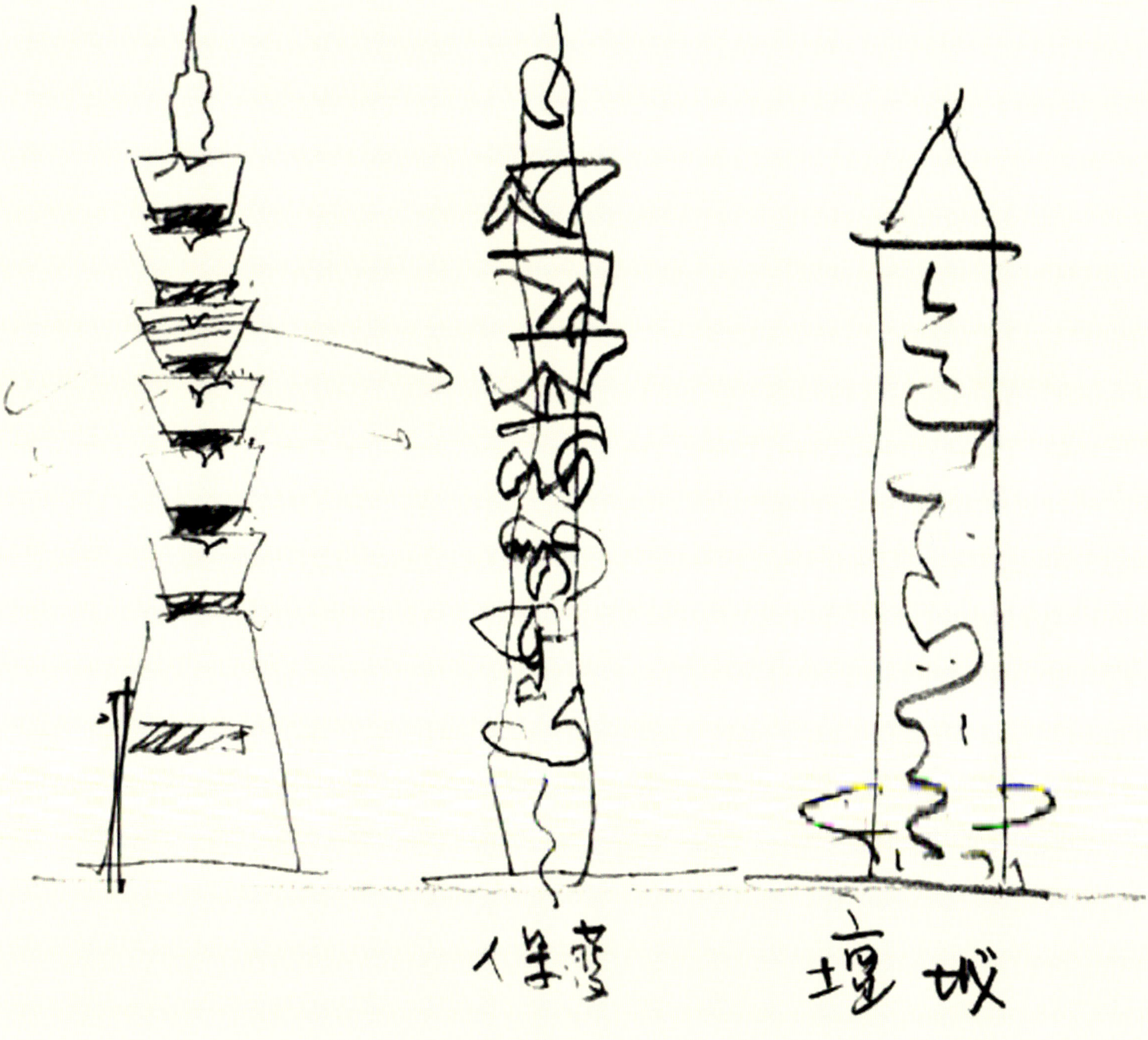
壇城

Since 1980, C.Y. Lee and C.P. Wang have run what has been for many years the leading architectural practice in Taiwan: C.Y. Lee & Partners Architects/Planners.

Their dual East-West early educational background and experience led C.Y. Lee and C.P. Wang to design contemporary architectural works that have their roots in traditional Chinese architecture, in their own revisited way. C.Y. Lee introduced Western technologies to his home country, where he pioneered high-rise architecture with a tall building in Kaohsiung in the late 1980s. As designer of the Grand 50 Tower (1992, 222 metres, 50 storeys), the T&C Tower (1998, 348 metres, 85 storeys) in Kaohsiung and later with TAIPEI 101, C.Y. Lee has been responsible for Taiwan becoming one of a select few countries that can proudly advertise themselves on the world map through state-of-the-art technological achievements of unprecedented height and technological challenges in a seismic region. While many tall projects designed by other architects in Chicago, Frankfurt or Hong Kong could interchange their location with each other, the projects designed by C.Y. Lee & Partners bear an unmistakable Chinese mark, expressing the cultural origin of their creators and the location for which they have been designed.

C.P. Wang, Architect, C.Y. Lee & Partners Architects/Planners.

Hung Kuo Building, C.Y. Lee & Partners, 1989, 19 storeys.

In Taipei, apart from TAIPEI 101, if one has to find a contemporary building that stands out from the mass, it will most probably be the Hung Kuo Building, developed by C.Y. Lee & Partners and Harace Lin. In contrast to many modern functional office buildings that are too often seen as a series of stacked slabs devoid of any detail or volume, this 19-storey building is truly tri-dimensional. Its exterior oriental sculptural details are as important as the different volumes of the interior, with its forest of majestic columns in the cavernous lobby and the upper atrium that serves as a rest space for the occupants. Both these spaces benefit from ample natural lighting from the vast windows that also act as lanterns at night when seen from afar. This project best summarises C.Y. Lee's and C.P. Wang's desire to design 'joyful' architecture that integrates contemporary reinterpretations of ancient Chinese cultural elements while proposing vast spaces for people to enjoy, rather than the cramped spaces that are often the case in Asia.

When C.Y. Lee starts talking about TAIPEI 101, he immediately points out three major issues that influenced his design for the tower: the view, the structural composition and safety concerns.

TAIPEI 101's inward slanting windows allow every occupant a more panoramic view over the city. The 7-degree inward slope of the window frames was influenced by the design of airport control towers with their inward-slanted observatories. But perhaps the most striking feature of TAIPEI 101 is the eight eight level stacked pods that make up the tower. The number eight represents prosperity and good fortune in Chinese culture; the effect of the stacked pods is a reinterpretation of the bamboo that 'swings in the air', explains C.Y. Lee. The third major aspect of the TAIPEI 101 design is safety - an integral part of the architecture and not just a mechanical additional feature. C.Y. Lee explains that the overall shape of each pod - each a reverse pagoda - creates

a terrace at the base of each of the eight stacked pods. The terrace levels allow occupants to switch from one staircase to the other if the need arises. In addition, each of these terrace levels that are used as mechanical levels also incorporate two refuge areas, a feature that allows the occupants who are not physically able to leave the building in case of emergency to await rescue in a secure area. Having explained the practical uses of the pod design, C.Y. Lee reminds us of the oriental silhouette he has created through the use of the reverse pagoda. While

Grand 50 Tower, Kaohsiung, Taiwan, C.Y. Lee & Partners, 1992, 222 metres, 50 storeys.

listening to him, we understand how important it is to design a project that reflects his own origins while at the same time perfectly integrates technology of both the Eastern and Western worlds - refuge levels, for example, are a safety amenity very common in China and not usually found in the USA or in Europe. TAIPEI 101 probably features more refuge levels than any other tall building.

C.Y. Lee believes that each country should have its own language when it comes to architecture and seems slightly concerned that few of his fellow architects of Asian origin share this view. He identifies three issues in relation to this aspect: the technology issue, the social issue and the cultural issue.

Regarding technology, C.Y. Lee has a clear and quick reply, 'You can buy it from the world', he says. However, we note that C.Y. Lee knows how to use technology in a dramatic way, transforming it into an actor that is part of the architectural drama.

'Equipment becomes public art', he says, referring to the 660-metric-ton tuned mass damper perched high above the tower and no longer seen as a pure engineering matter. C.Y. Lee thinks a bar should be designed around the tuned mass damper 'so people don't know if it is the ball moving, or if they are moving', he adds with good humour. 'Changing machines into art' is his motto. Indeed, the tuned mass damper is of great interest to visitors to the TAIPEI 101 observatory.

C.Y. Lee also uses technology for its environmental role: steel was chosen because it can be recycled; the double low-E glass façade was selected because it reduces heat by two-thirds; grey water is used to flush toilets and cool the building. Regarding the social issue, C.Y. Lee says that high-rise buildings are becoming the central life, the place to go in the city; that towers should be iconic and that a social plaza should be implemented in such projects. In TAIPEI 101, the covered landscaped urban City Square on the fourth level of the tower, above the shopping mall, is the social plaza where people gather for a drink, lunch or just for reading a book on a bench. 'The cultural issue is the most important one', insists C.Y. Lee. He reminds us that after 5000 years of Chinese history and the wooden structures that disappeared with the 1966 Cultural Revolution, it was time to start a new architectural system for the region. He stresses that it is the responsibility of the architect to develop his own language, adapted to the region where he is active.

This approach can be observed in TAIPEI 101 at all levels. The overall oriental allusion to bamboo gives the project an organic, ever-growing, resistant and flexible aspect. Chinese cultural elements can be found throughout TAIPEI 101 from base to top. In addition, artworks by a variety of international artists are prominently displayed throughout the project. In the hands of C.Y. Lee, the integration of culture and technology has produced a building unlike any other.

When asked if he expects his experience of having built the world's tallest building to lead him to build structures of this height in other places of the world, C.Y. Lee modestly implies that he wouldn't feel very comfortable working in other regions (besides Asia). This sensibility has indeed led C.Y. Lee to work in several locations in China, including Shanghai, where he recently completed the renovation of a group of contemporary buildings. These poor-quality buildings were erected about 15 years ago at the eastern edge of the Bund, the main Art Deco area in the city. The Zhong Shan (East) Road renovation included re-cladding the façades of a high-rise tower and adding new penthouse levels, reshaping and transforming a brutal element into a friendlier urban piece of architecture and complying with both the urban fabric and aesthetics and the local history, an exercise that is dear to C.Y. Lee.

Harace Lin reminds us that coincidences are part of the process of creating imaginative achievements. But perhaps it is no coincidence that the Asian architect who has designed the most recognisable high-rise project in the world is also the architect who worked in the 1970s with William L. Pereira, the American architect who designed some of the most recognisable high-rises in any skyline, including the 48-storey Transamerica Pyramid in San Francisco in 1972. Some architects were born to give a lasting identity to San Francisco, others to Taipei.

Zhong Shan (East) Road renovation project, Shanghai, C.Y. Lee & Partners, 2007.

CULTURAL INFLUENCES

A combination of traditional Chinese elements and high technology create TAIPEI 101's graceful, soaring profile. Design features borrow heavily from Chinese culture, including aspects of numerology, nature, Feng Shui, traditional architecture and symbology.

An oriental silhouette is created through the use of the reverse pagoda.

2009
2009

Nature · The oriental allusion to bamboo and its overall organic form give the project an ever-growing, resistant and flexible appearance.

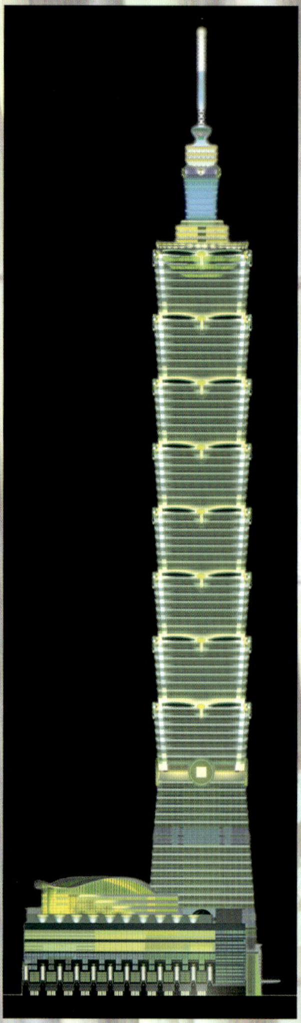

The pagoda form reflects the beauty of oriental flowers.

Numerology · The concept is based on the number eight, a numeral representing prosperity in Chinese culture. The design is based on eight-floor interlocking structural units, connected one by one on top of each other to form the whole. This rhythmic aesthetic is new to skyscraper design.

The architects explored a number of designs before settling on the idea of dividing the building into eight sections of eight floors each - the sections are clearly delineated on the exterior of the building.

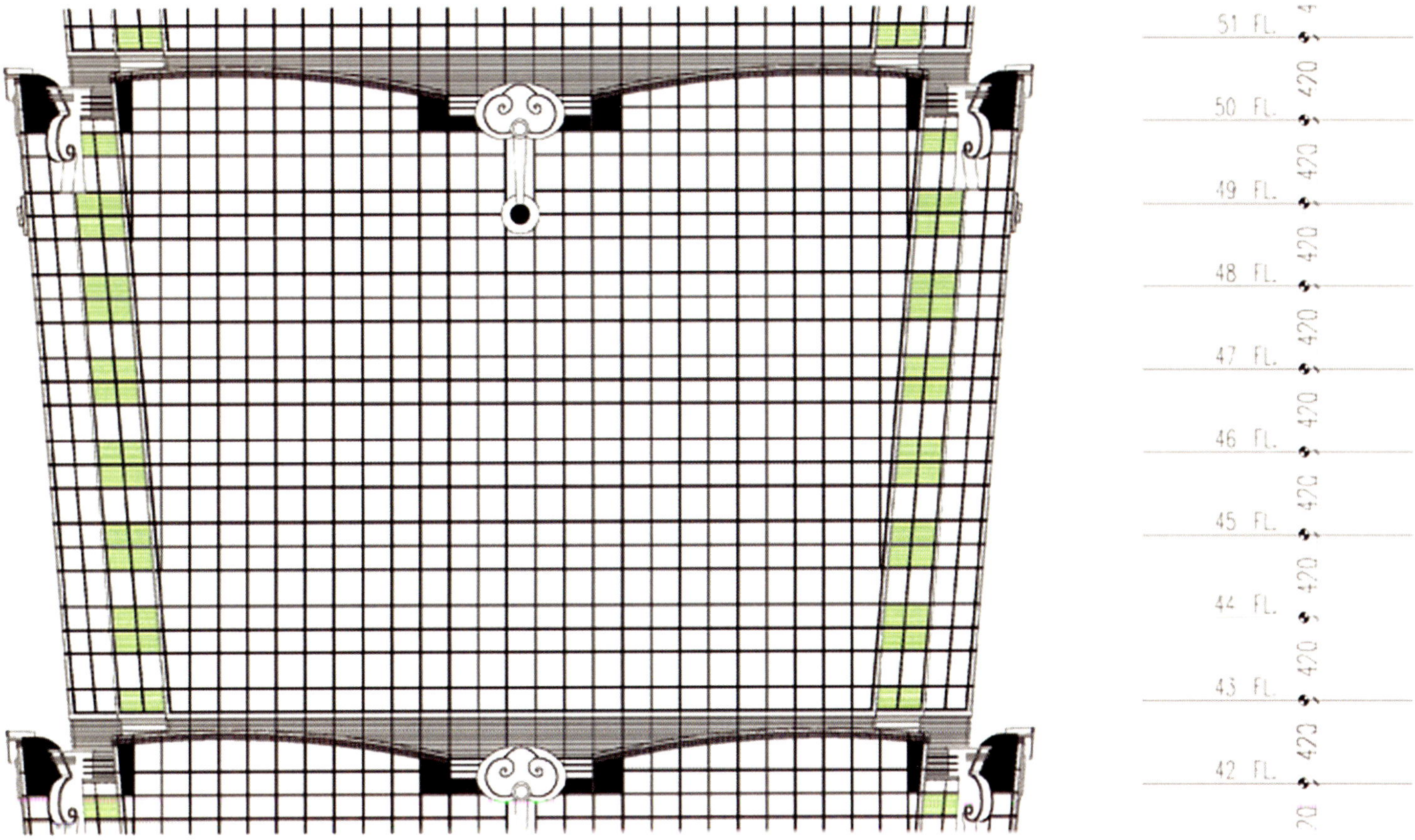
51 FL.
420
50 FL.
420
49 FL.
420
48 FL.
420
47 FL.
420
46 FL.
420
45 FL.
420
44 FL.
420
43 FL.
420
42 FL.

FENG

Feng Shui

Feng Shui, literally translated as 'wind and water', refers to the ancient Chinese practice of arranging space to achieve harmony with the environment. A Feng Shui master was asked to review the design before it was finalised; not much had to be changed although the master insisted that the colour of the tower be very light.

SHUI

Decorative elements

Chinese cultural elements can be found throughout TAIPEI 101 from base to top. For instance, at the top four corners of each of the eight pods are Xi Qi Yang Yang in the form of silver decorated elements symbolising auspicious blessing. These elements represent a contemporary reinterpretation of the dragon and stone drums that are found in traditional Chinese temple architecture.

On each of the eight segments of the tower façade and inside the mall are a series of metallic silver clouds symbolising the Ruyi 'as you wish' pendants worn in ancient China as a symbol of luck and happiness. The Ruyi are also found in the form of the roof above the mall.

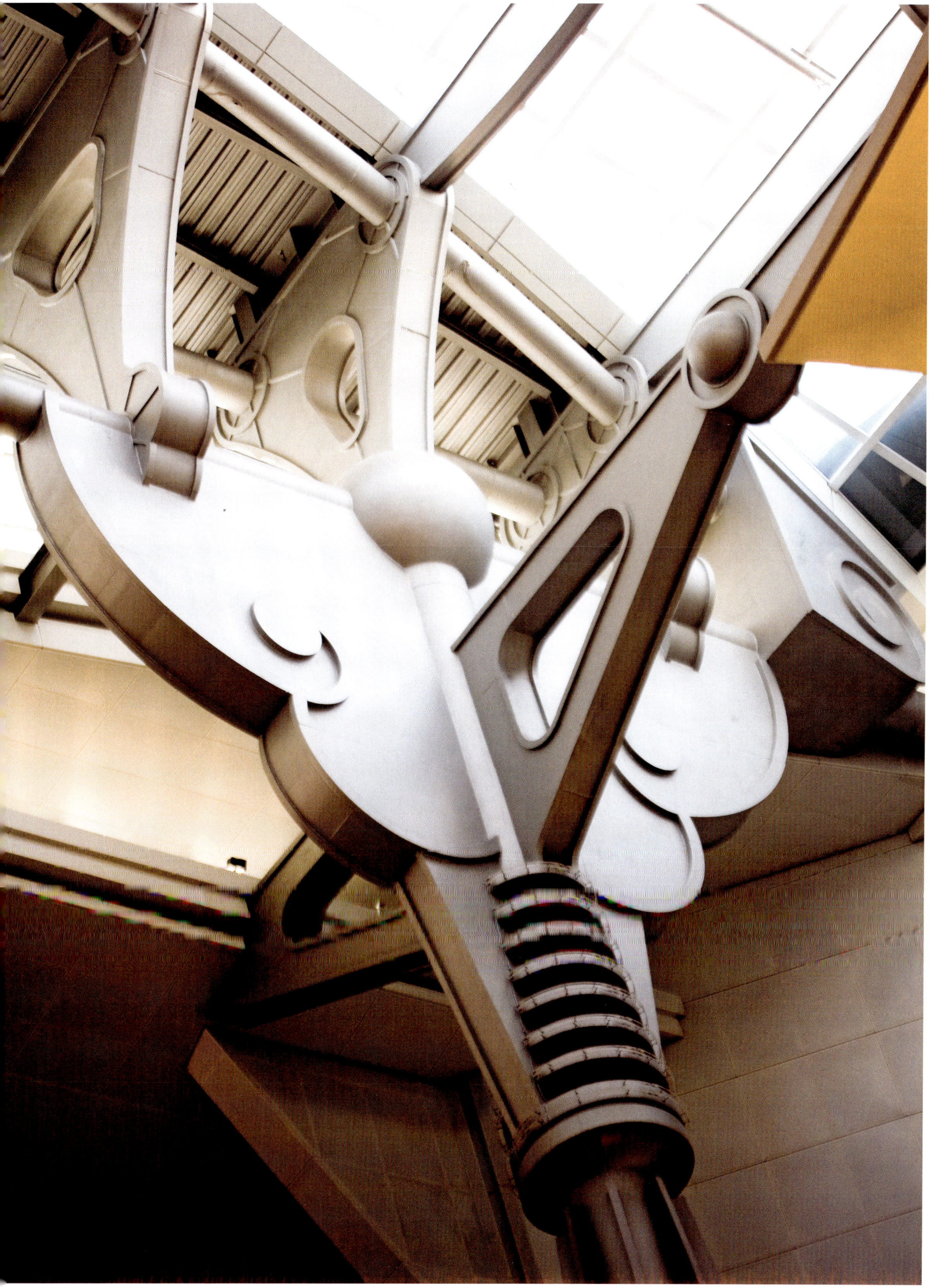

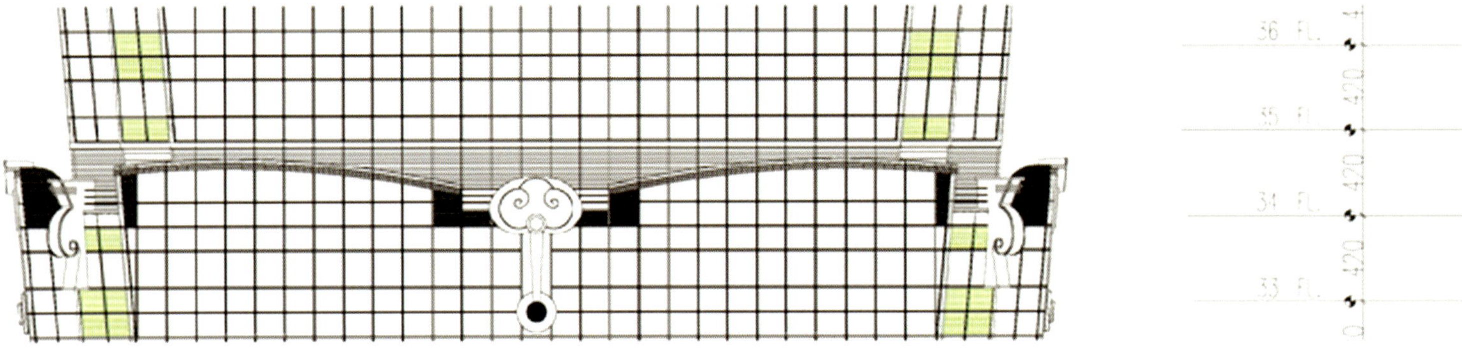
36 FL.
35 FL.
34 FL.
33 FL.
420
420
420

TAIPEI 101

To symbolise prosperity and abundance, four huge ancient Chinese coins revisited in a contemporary design stand on the four faces of the lower level of the first of the eight pods. These eight pods look like a series of stacked Chinese ingots used in ancient times as money. The TAIPEI 101 logo itself symbolises three interlocking ancient Chinese coins, each symbolising in turn the three TAIPEI 101 main units: the office tower, the shopping mall and the observatory.

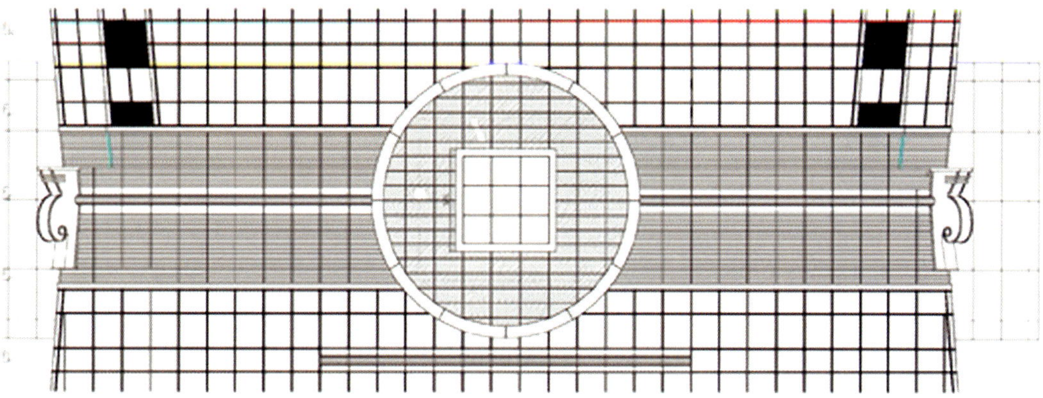

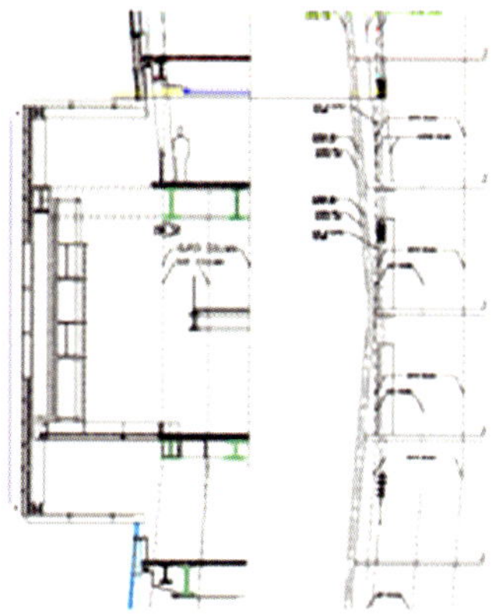

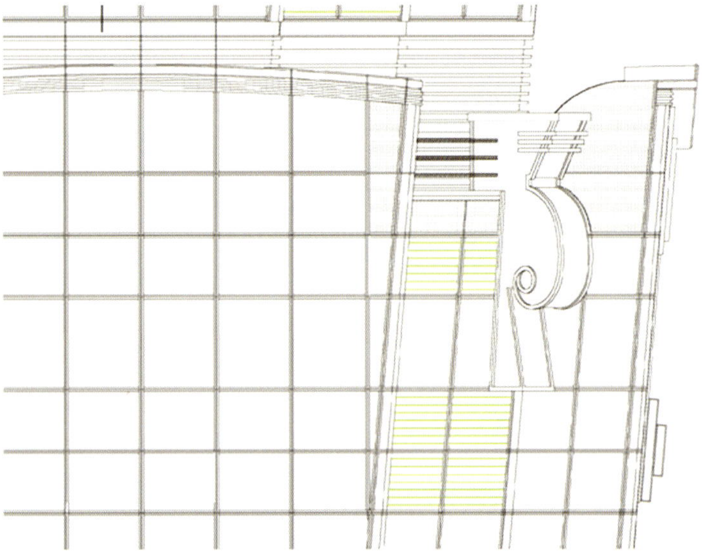

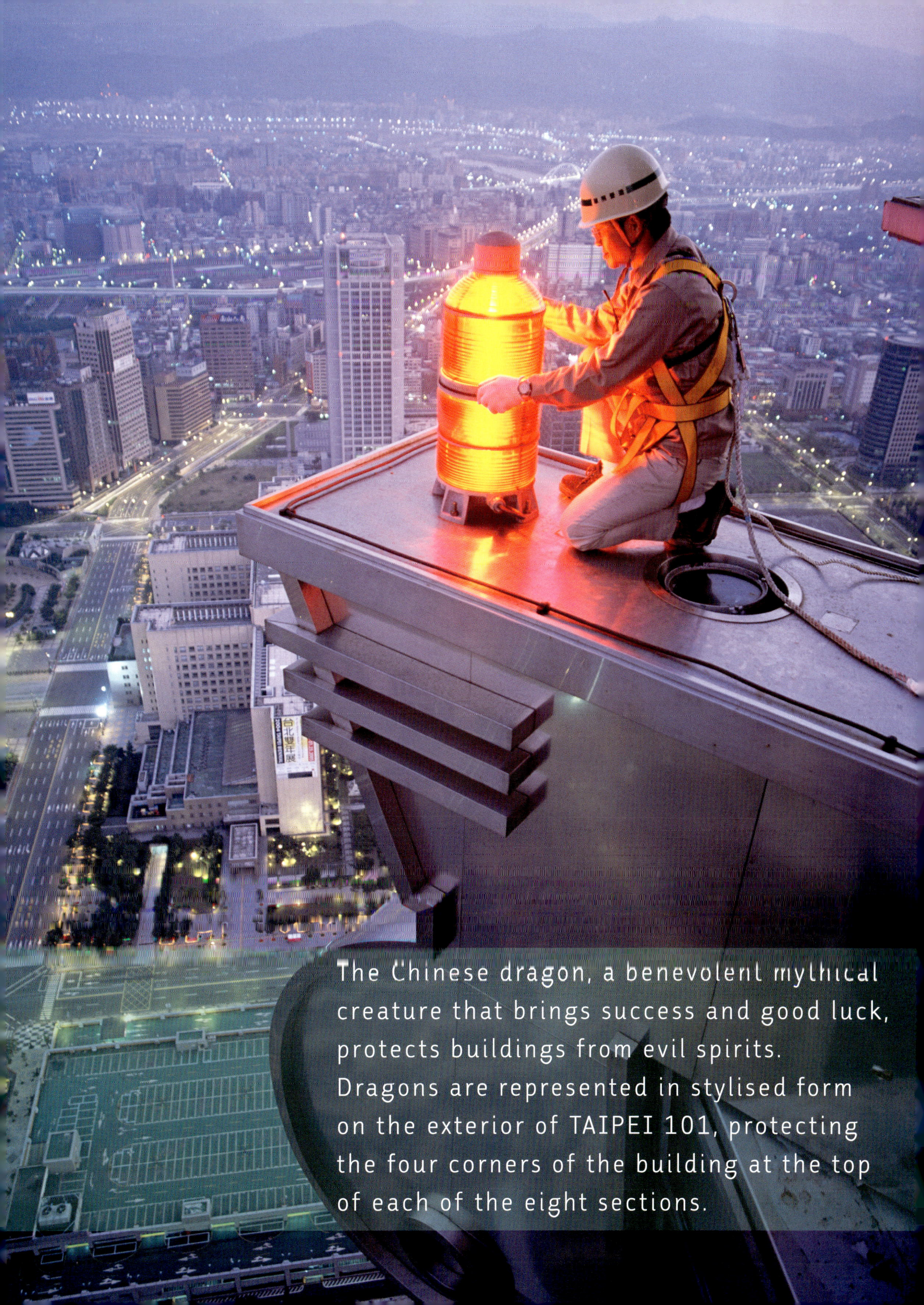

The Chinese dragon, a benevolent mythical creature that brings success and good luck, protects buildings from evil spirits. Dragons are represented in stylised form on the exterior of TAIPEI 101, protecting the four corners of the building at the top of each of the eight sections.

DES

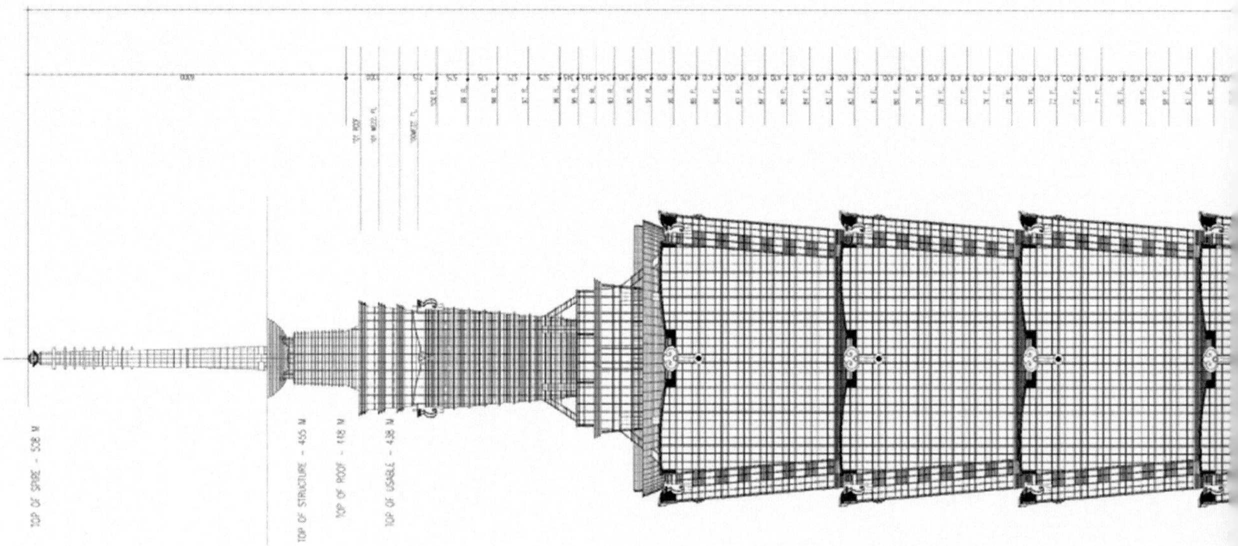
TOP OF SPIRE - 508 M
TOP OF STRUCTURE - 455 M
TOP OF ROOF - 448 M
TOP OF USABLE - 438 M

CON

IGN

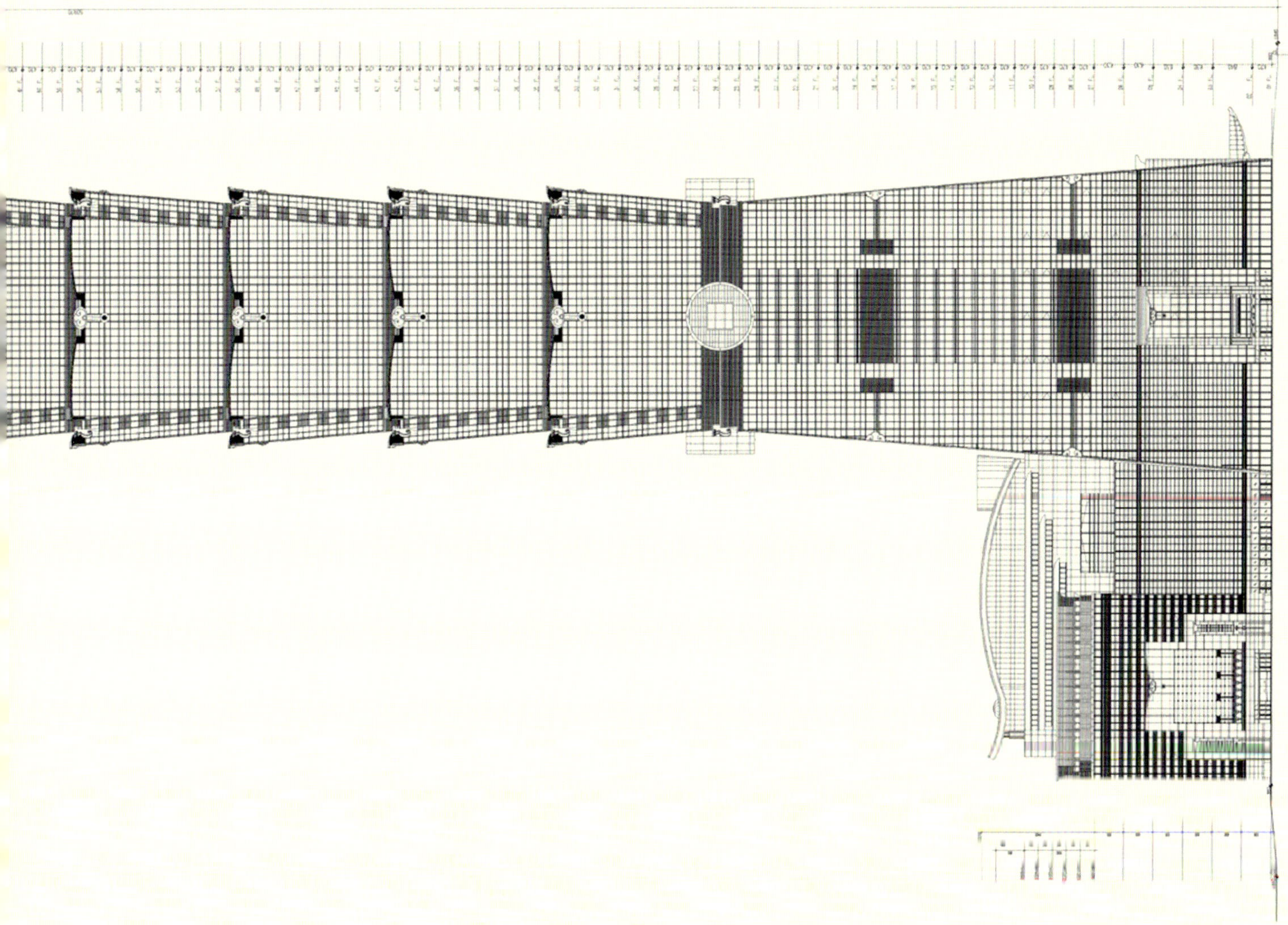

CEPT

The design concept is based on the number eight, a numeral representing prosperity in Chinese culture. The design is based on eight-floor interlocking structural units, a rhythmic aesthetic new to skyscraper design. Inclining 7 degrees inwards, the floor plate increases in size as it gets higher within each section.

The 200,000-square-metre tower rises 101 storeys from a footprint of 4900 square metres. Eight stacked dimensionally identical modules, each eight storeys, sit above a lower, 25-storey section with an inward slope of 5 degrees. The modular arrangement creates external fire safety decks at the base of each eight-floor module. Each module is isolated by smoke and fire barriers and contains independent security systems.

At the top are 11 mechanical/communications levels, which step back three times. A 60-metre-tall pinnacle rising from the 101st storey completes the tower at 508 metres. A shopping mall is in the linked five-storey podium at the base of the tower.

Wind effects usually created on the surface of high-rise buildings are minimised by the building's zigzag W-shaped corners, a feature designed after wind tunnel testing.

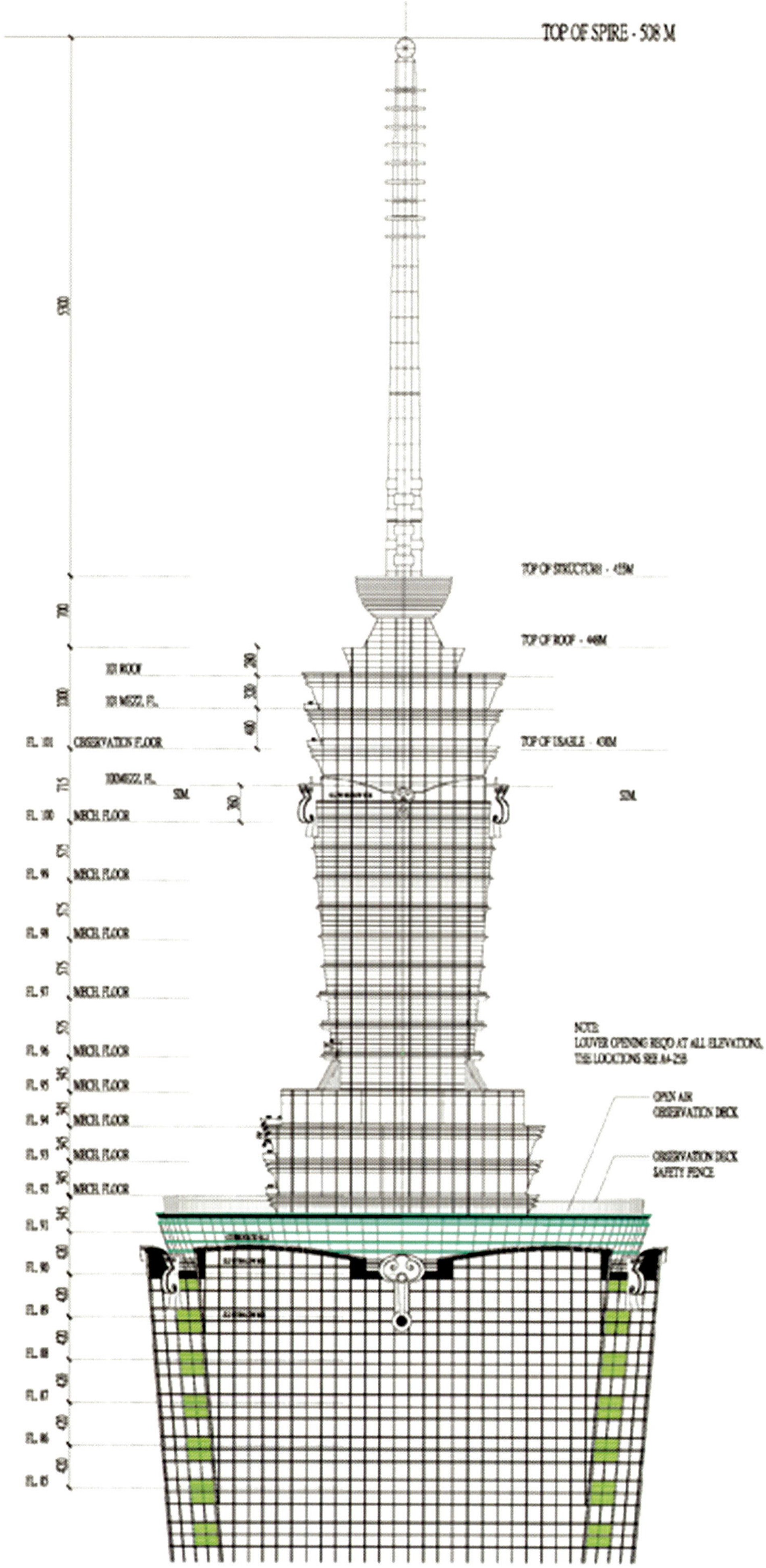
TOP OF SPIRE - 508 M

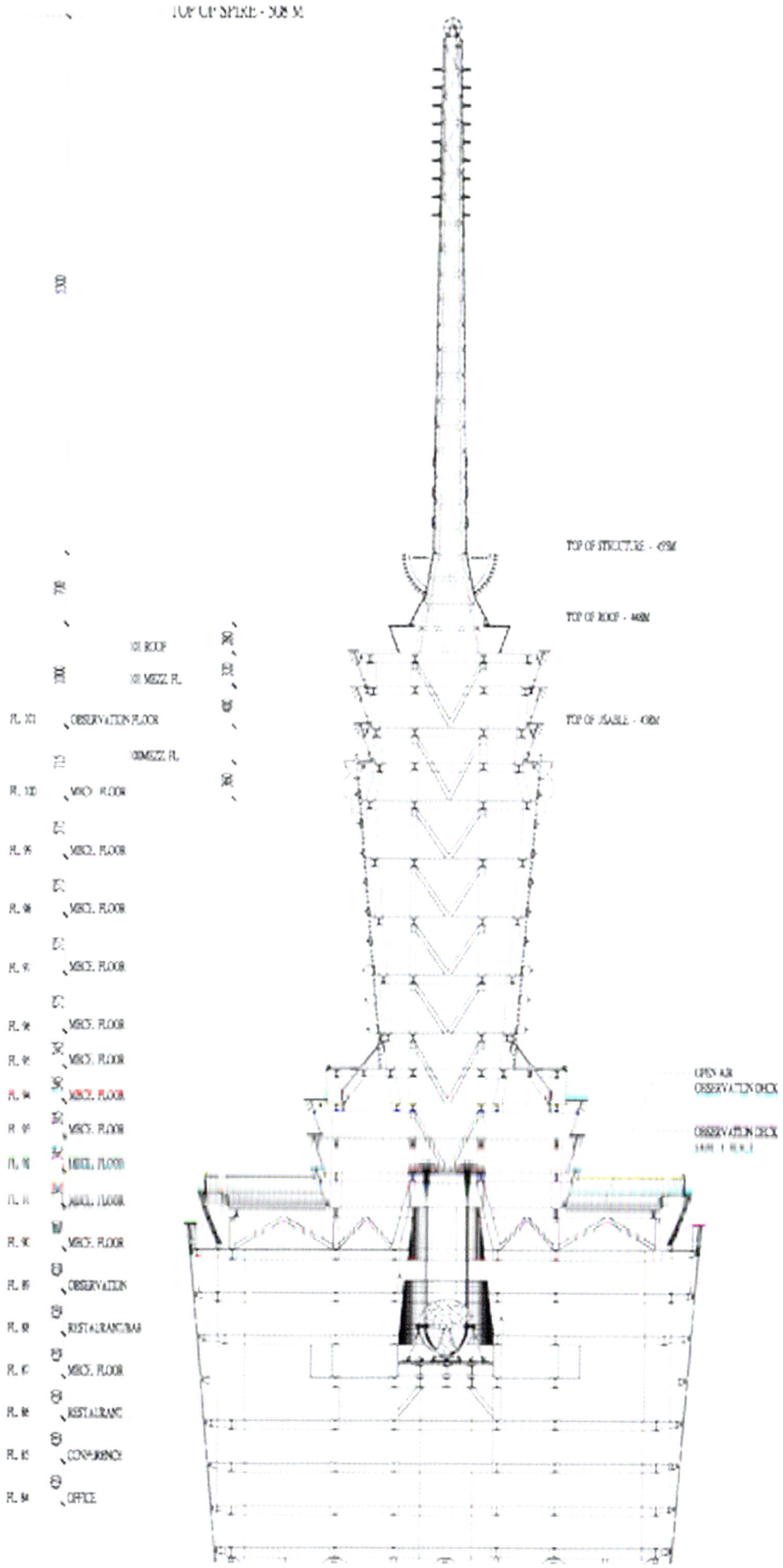
OBSERVATION FLOOR
OBSERVATION
RESTAURANT
OFFICE

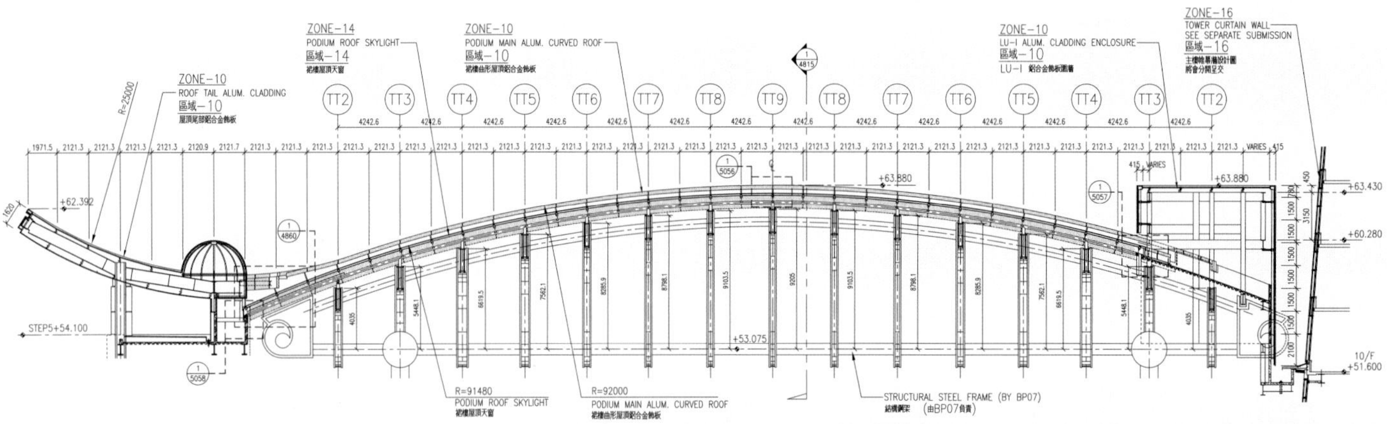

裙樓,屋頂天窗位置垂直剖面圖

1/4814 LONGITUDINAL SECTION THRU ROOF SKYLIGHT

AT ZONE-10,14

區域10,14

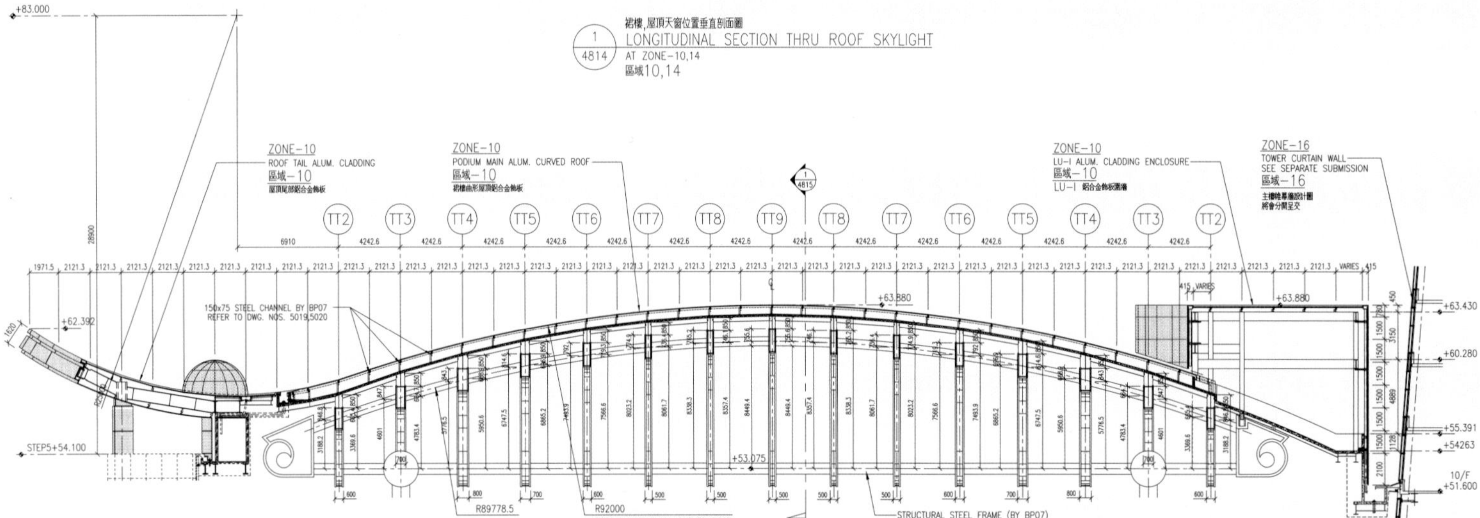

裙樓,曲形屋頂位置垂直剖面圖

2/4814 LONGITUDINAL SECTION THRU CURVED ROOF

AT ZONE-14

區域14

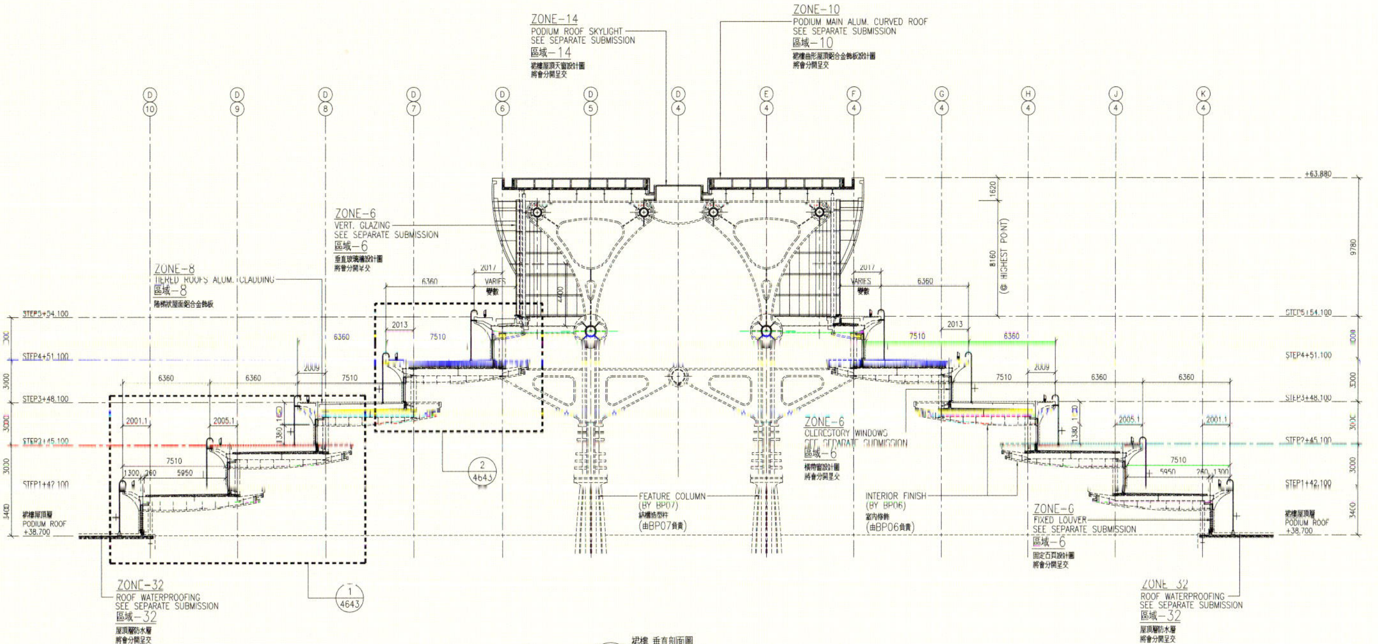

裙樓,垂直剖面圖
1
4642
VERT. SECTION AT PODIUM
PARENT DWG. NO. 4601, 4602, 4603, 4604, 4605, 4606, 8609
(SCALE = 1 : 150)
參閱平面圖號: 4601, 4602, 4603, 4604, 4605, 4606, 8609
(比例= 1 : 150)

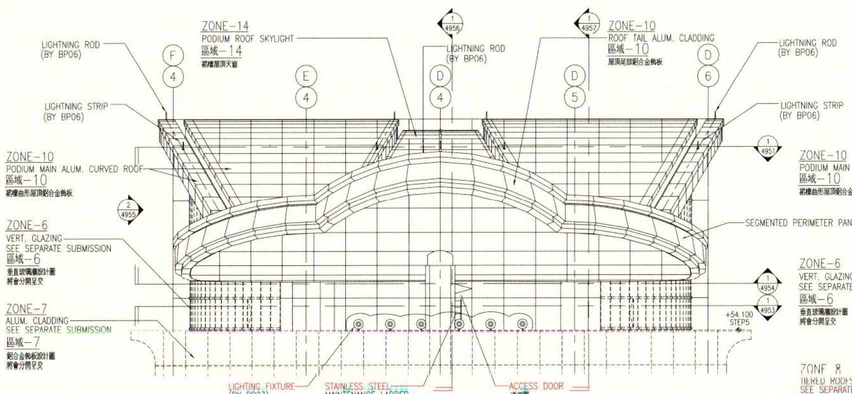

裙樓,局部放大立面圖
1
4955
BLOW UP ELEVATION AT ROOF TAIL
PARENT DWG. NO. : 4951

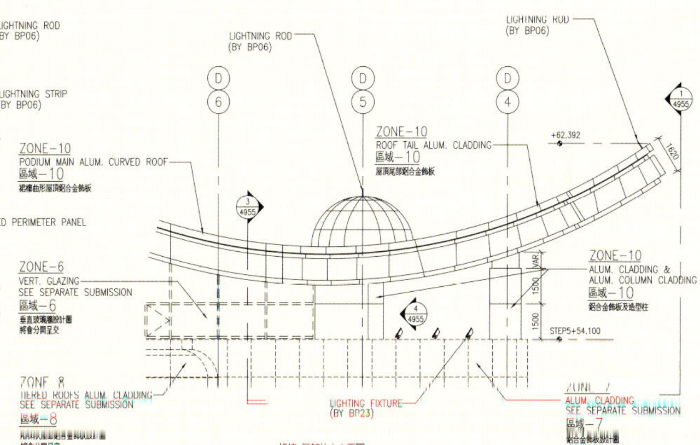

裙樓,局部放大立面圖
2
4955
BLOW UP ELEVATION AT ROOF TAIL
參閱平面圖號: 4951

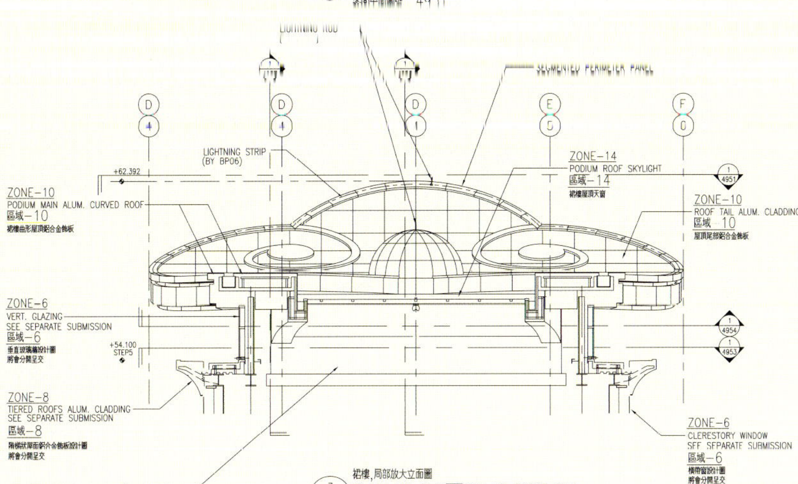

裙樓,局部放大立面圖
3
4955
BLOW UP ELEVATION AT ROOF TAIL
PARENT DWG. NO. : 4951, 4954, 4955
參閱平面圖號: 4951, 4954, 4955

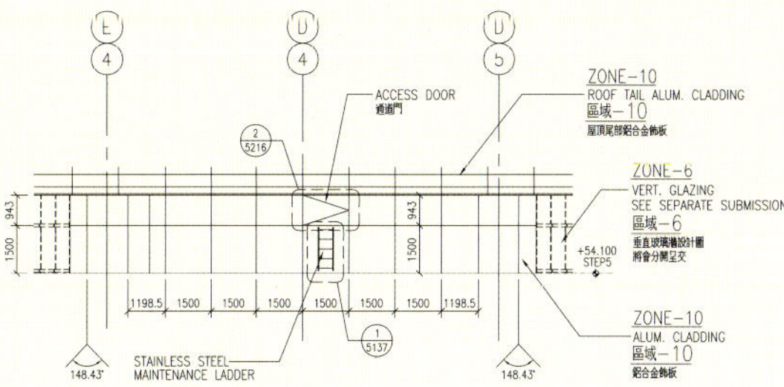

裙樓,局部放大立面圖
4
4955
BLOW UP ELEVATION AT ROOF TAIL
PARENT DWG. NO. : 4953
參閱平面圖號: 4953

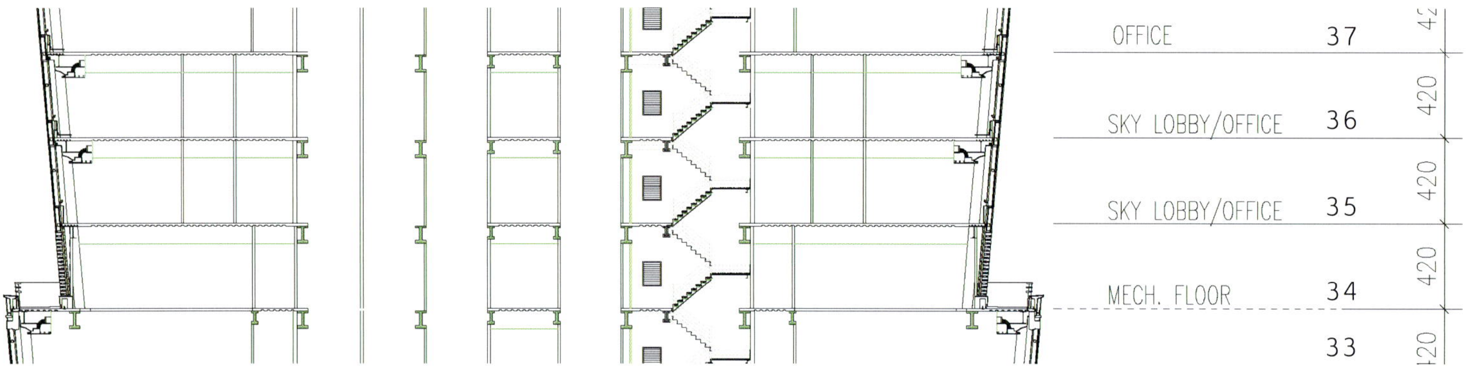
OFFICE
37
SKY LOBBY/OFFICE
36
SKY LOBBY/OFFICE
35
MECH. FLOOR
34
33
420
420
420

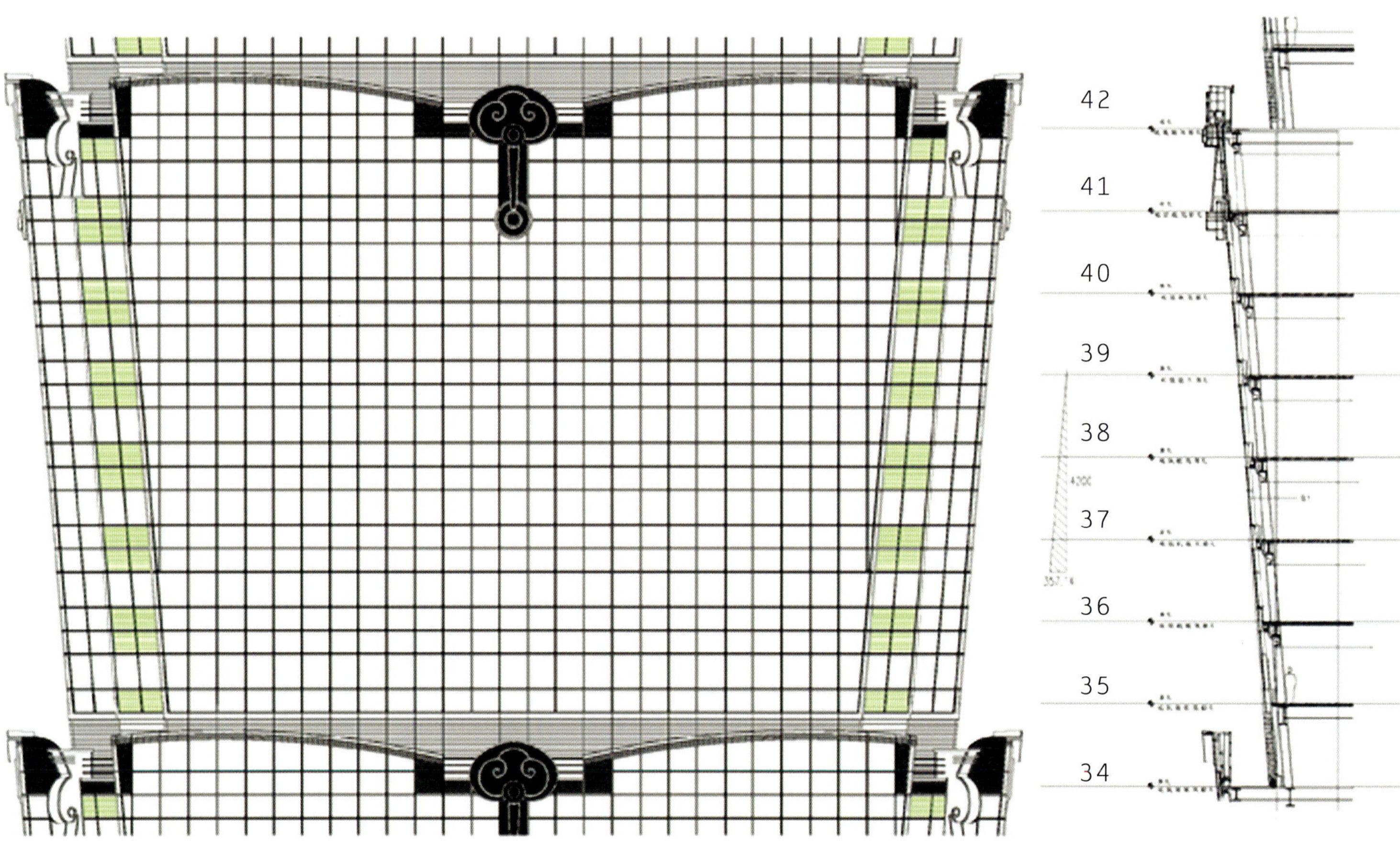
42
41
40
39
38
37
36
35
34

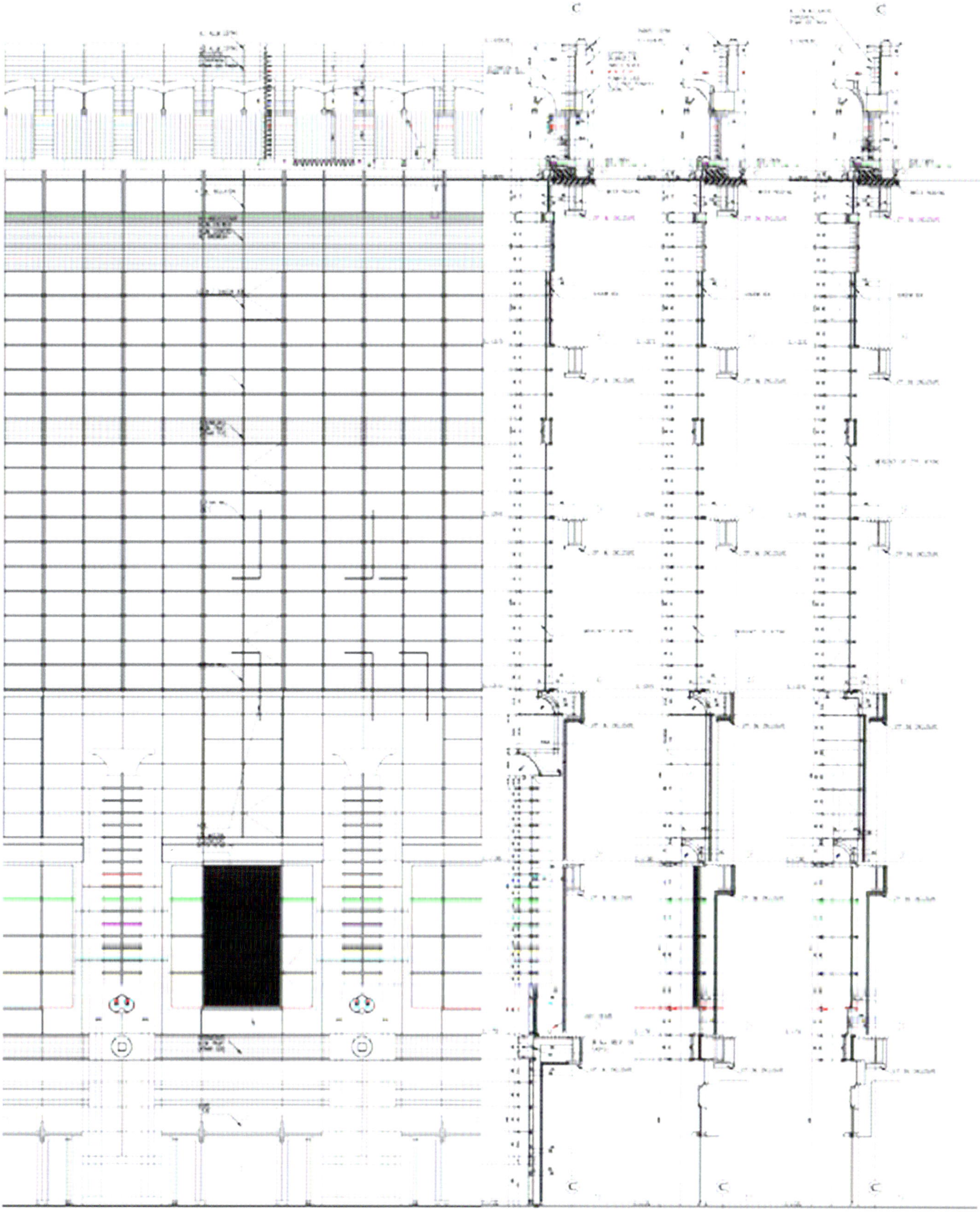

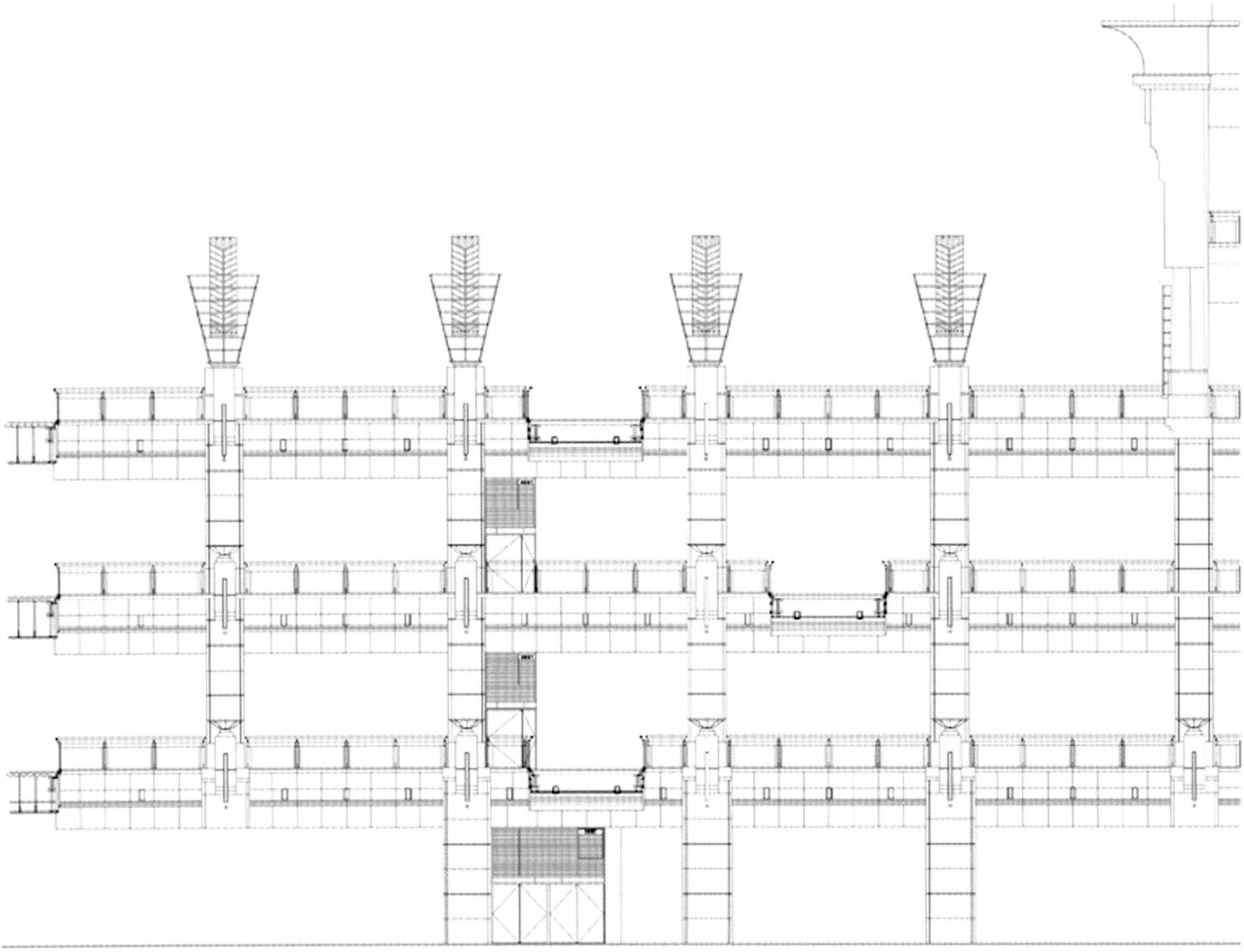

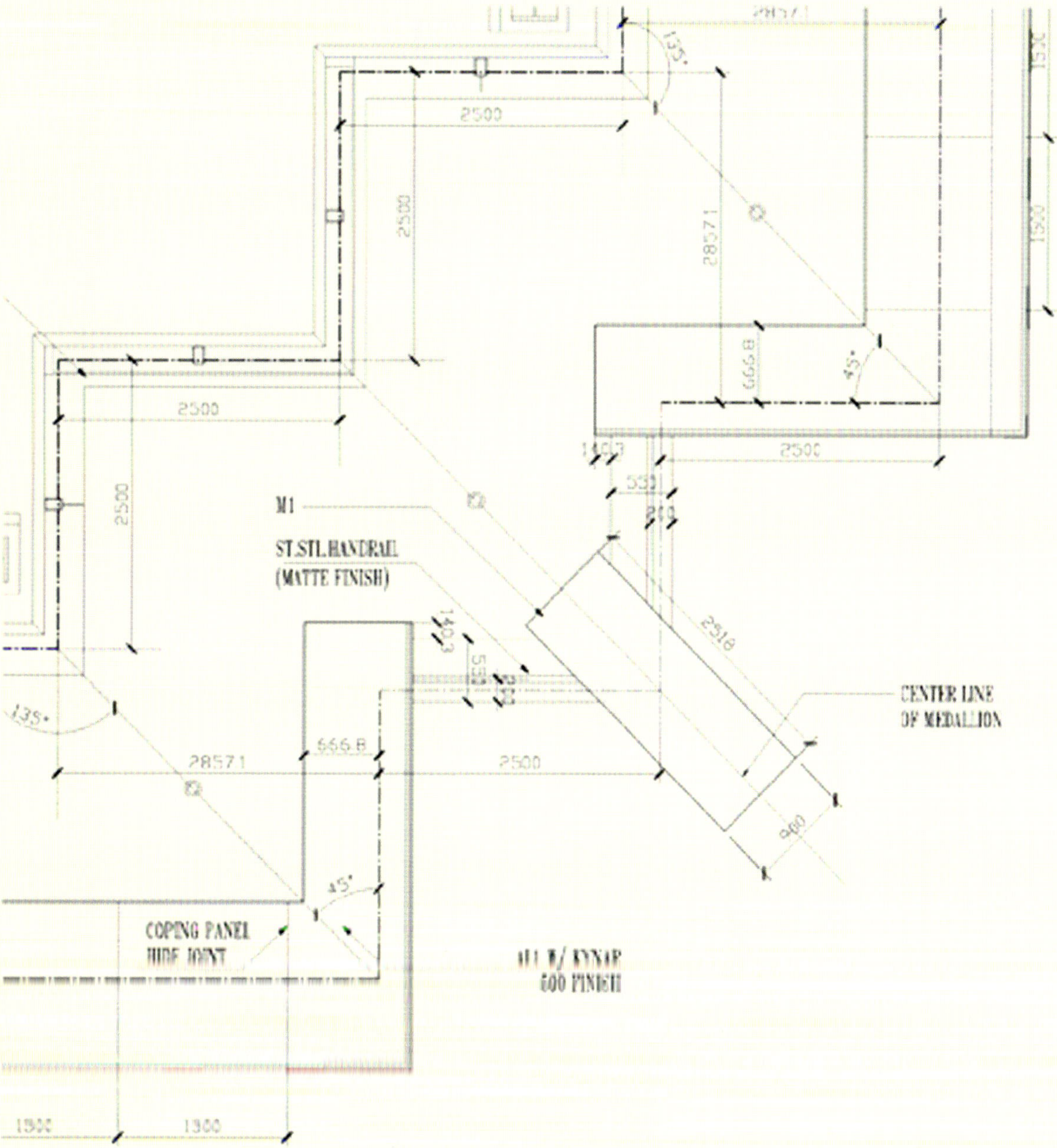

M1
ST.STL.HANDRAIL
(MATTE FINISH)
CENTER LINE
OF MEDALLION
COPING PANEL
HIDE JOINT

STRUCTURAL DESIGN

In plan, the steel cruciform structure of TAIPEI 101 is reminiscent of the concrete megastructure designed in 1989 for an unbuilt 125-storey tower in Chicago (developed by Miglin-Beitler and designed by Cesar Pelli & Associates and Thornton Tomasetti), which was scheduled to become the next world's tallest building at a height of 609 metres. The perimeter bracing approach was used in many of the super-tall buildings built since the 1960s, including the World Trade Center in New York, which was made of a steel-framed tube structure; the 100-storey John Hancock Center in Chicago, which is made of a steel braced tube structure; and the Sears Tower, also in Chicago, which employs a steel bundled tube structure.

While the sloping façades of TAIPEI 101 did not present the ideal situation for perimeter framing, they did present an opportunity to the engineers to incorporate state-of-the-art technology into the building. Structurally, the tower is a working showcase of the most advanced building technology, with a structural steel megastructure design that provides maximum strength and stability.
The core and outrigger scheme consists of 11 mega sections, each comprising eight storeys. High-strength steel box columns filled with high-strength reinforced concrete form a strong framework, with eight megacolumns as backbones. Two megacolumns, usually 3 metres by 2.4 metres, are located along each side of the building, 22.5 metres apart and almost touching the sloping perimeter wall at the base of each eight-level section. Main floor girders connect each megacolumn through moment connections with a core corner column. The 22.5-metre-square core comprises 16 box columns in four lines, which are generally braced between floors.

To maximise safety, the 8-centimetre-thick megacolumns are filled with 10,000-pounds-per-square-inch reinforced concrete up to level 62, far exceeding the typical standard of 6000 pounds per

The tower is built on 380 concrete piles, each 1.5 metres in diameter, sunk 80 metres into the ground.

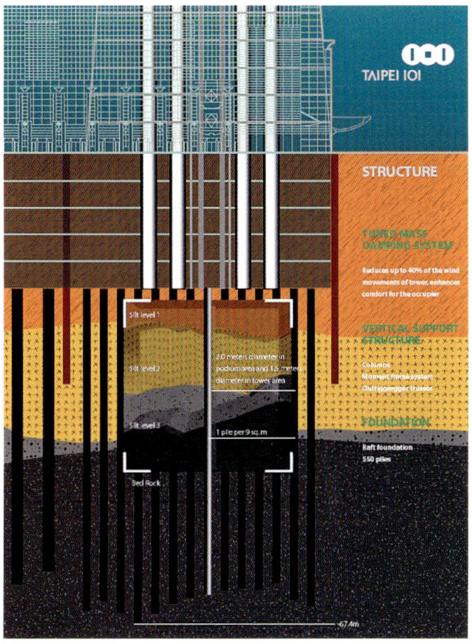

square inch. At each mechanical level – every eighth level and also used as refuge levels – massive steel outrigger trusses connect the megacolumns to the core. 'If this building is going to be damaged, other buildings in Taipei will collapse', said Shaw-song Shieh, Chairman of Evergreen Consulting Engineering, Inc. (*ENR* November 24, 2003). Because the building is subject to typhoons and earthquakes, the structural challenge was to allow it to be simultaneously stiff – to withstand typhoons – and soft – to resist earthquakes.

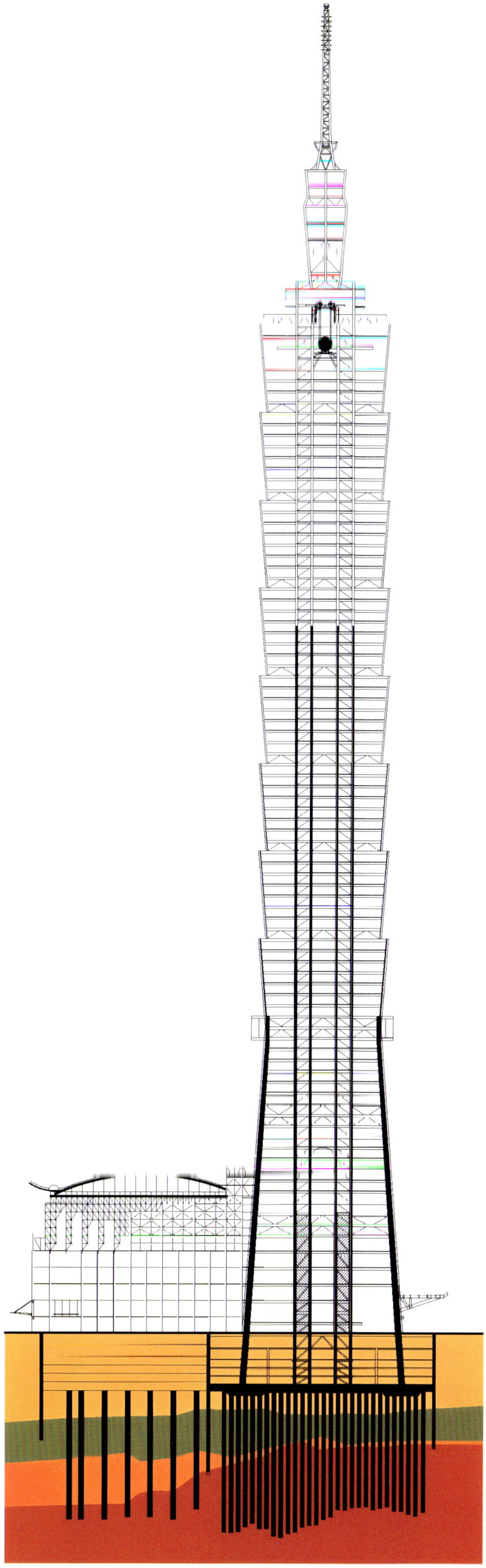

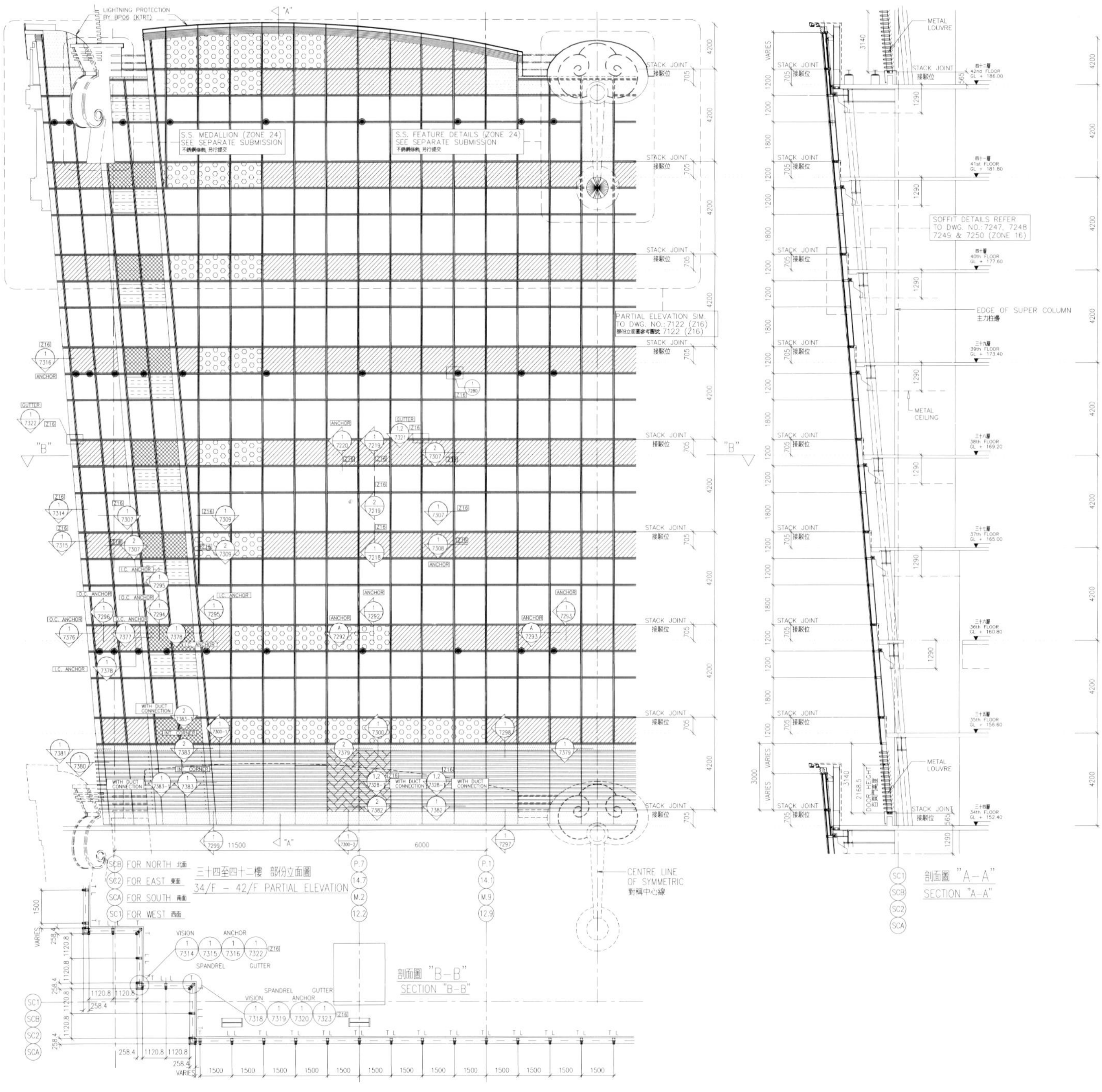

LIGHTNING PROTECTION BY BPD6 (KTRT)
S.S. MEDALLION (ZONE 24) SEE SEPARATE SUBMISSION
S.S. FEATURE DETAILS (ZONE 24) SEE SEPARATE SUBMISSION
STACK JOINT 接駁位
PARTIAL ELEVATION SIM. TO DWG. NO. 7122 (Z16)
SOFFIT DETAILS REFER TO DWG. NO. 7247, 7248 7245 & 7250 (ZONE 16)
EDGE OF SUPER COLUMN 主力柱邊
METAL LOUVRE
METAL CEILING
ANCHOR
GUTTER
WITH DUCT CONNECTION
SCB FOR NORTH 北面
SC2 FOR EAST 東面
SCA FOR SOUTH 南面
SC1 FOR WEST 西面
三十四至四十二樓 部份立面圖
34/F – 42/F PARTIAL ELEVATION
CENTRE LINE OF SYMMETRIC 對稱中心線
VISION
SPANDREL
剖面圖 "B–B"
SECTION "B–B"
剖面圖 "A–A"
SECTION "A–A"

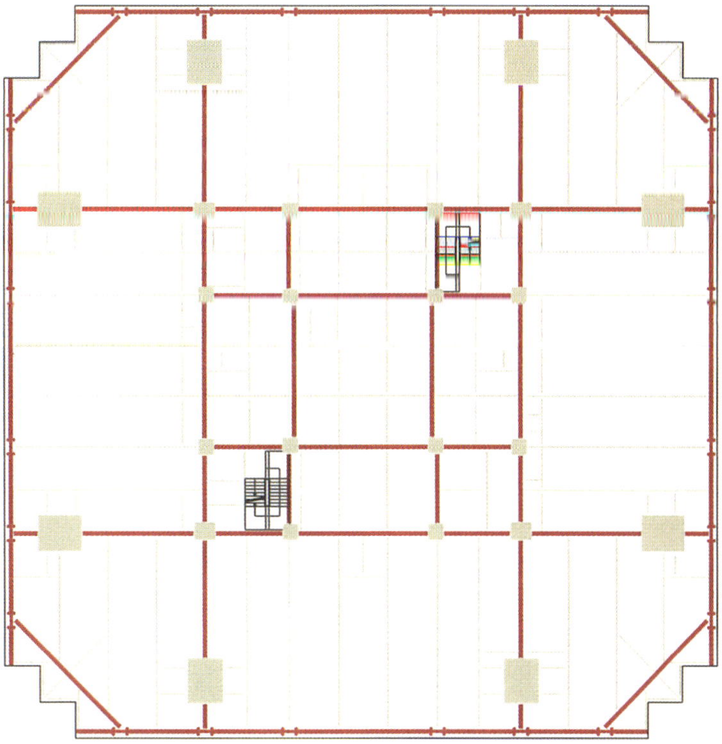

The structure is reinforced by a moment frame system linking the columns on all floors.

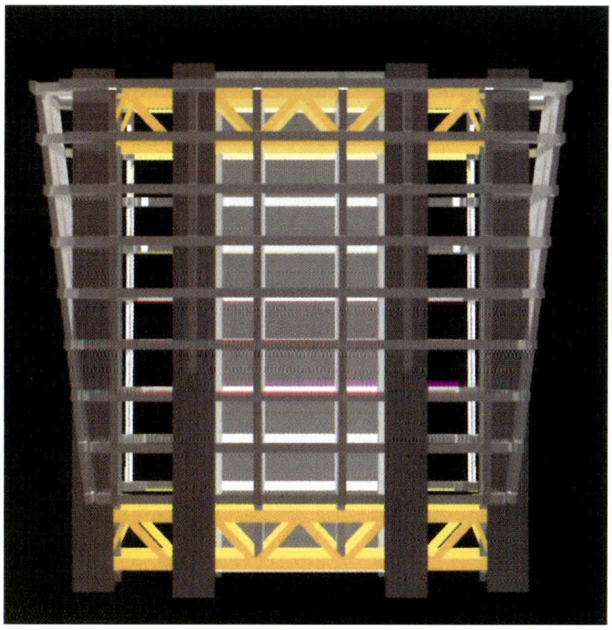

Massive steel outrigger trusses span between the columns on every eight floors.

36 columns provide vertical support, including eight megacolumns around the perimeter.

Curtainwall

The unforgettable blue-green façade of TAIPEI 101 not only contributes to the building's iconic appearance, but is a further example of technological innovation, protecting the occupants of the building against typhoons and earthquakes, while providing impressive views.

The tower façade has an area of 120,000 square metres and is made of pressure-compensated elements that are designed to withstand earthquakes and typhoon wind loads with wind speeds of up to 250 kilometres per hour. The 16,000 façade elements of the tower are 4.1 metres high and 1.5 metres wide and consist primarily of glazing, rustproof steel and aluminium.

The actual elements of the tower façade are repeated frequently. However, the elements are distinguished by additional external details such as 32 stainless steel ornaments that are two storeys high and each weigh 5 tonnes. A stainless steel ornament (coin) with a diameter of 14 metres and a depth of 4 metres was installed on every side of the building on the 25th storey. This ornamentation consists of prefabricated glass and aluminium elements, and is supplemented by suspended cladding.

The 4000 façade elements of the podium were made of glass, sheet metal and natural stone. The podium roof is made of curved steel and glass.

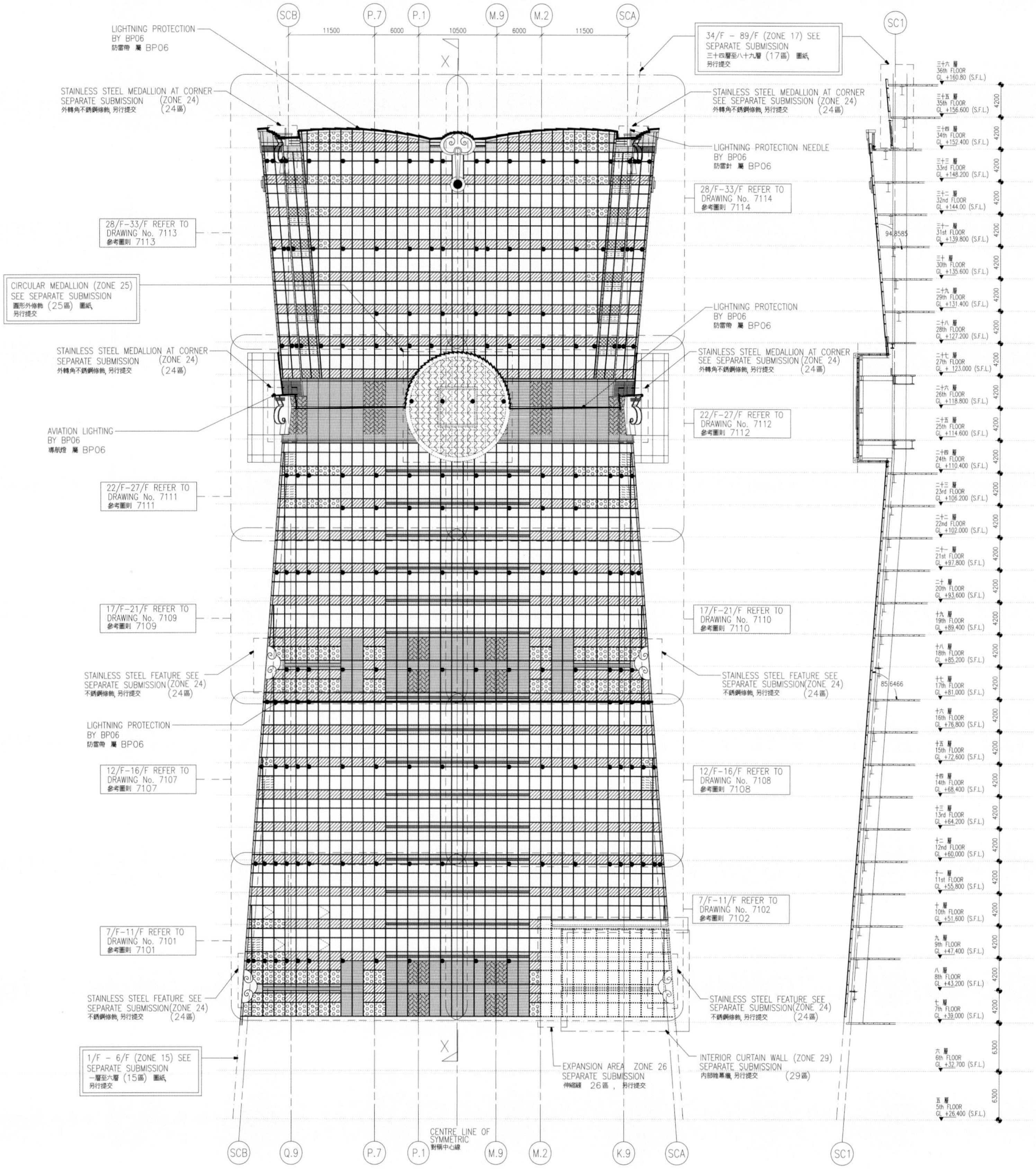

北面 "A" 立面圖 (主樓)

NORTH ELEVATION "A" AT TOWER (ZONE 16)

剖面圖 X – X

SECTION X – X

VERTICAL TRANSPORTATION

Transfer floor lift lobby (floor 36)

Vertical transportation, which includes elevators, escalators and a sky lobby system, plays a major role in ensuring an efficient working and shopping environment within TAIPEI 101.

TAIPEI 101's two pressurised observation elevators are the fastest in the world. They ascend at 1010 metres per minute and decend at 600 metres per minute. This speed means that a trip from the 5th floor to the 89th floor observatory, a distance of 382.2 metres, takes an astounding 37 seconds.

These elevators incorporate the world's first pressure control system, which adjusts the atmospheric pressure inside a car by using suction and discharge blowers mounted on each car top, preventing those riding inside the car from experiencing the discomfort of 'ear popping'. An active control system cancels vibrations by moving the counter mass in the opposite direction, based on vibration data collected by a sensor installed in the car. The whistling noise usually produced by a car running at a high speed inside a narrow hoistway is minimised to ensure a pleasant trip, with the minimal perception of movement, to the top of the building. Onboard digital speedometers are continuously displayed so that passengers can monitor their progress up the building.

TAIPEI 101's 34 double-deck elevators, combined with two double-level sky-lobbies, allow an economical and efficient use of the building. The aim is that the wait for an elevator should be no longer than 30 seconds. Apart from the mall portion in the lower levels (B1 to 5F) and the upper communication levels (92F to 101F), the main body of the office tower is divided into three parts: the low zone, the mid zone and the high zone. The three parts are separated by two two-level sky-lobbies found at levels 35-36 and levels 59-60. People travelling to a level between levels 37 and 57 will first take an express shuttle elevator to sky-lobby level 35-36 and move to a local elevator to reach levels 37 to 57. Occupants of and visitors to levels 61 to 88 will first take an express shuttle elevator to

Model of the observatory shuttle elevator, on display on the 89th floor observatory.

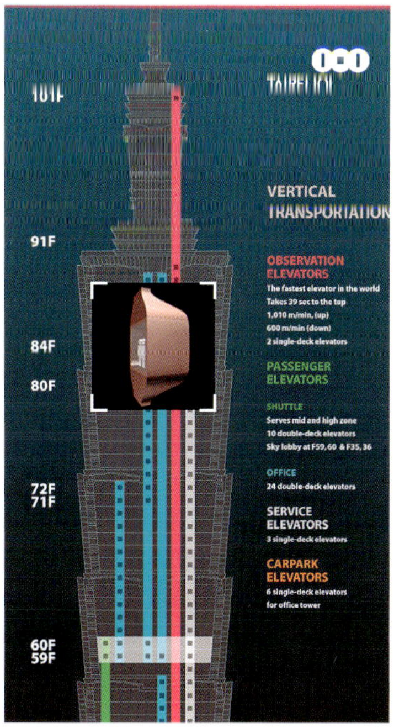

Special effects on the ceiling of the observatory shuttle elevator.

sky-lobby level 59-60 and then take a local elevator to levels 61 to 88. Local elevators in the main lobby are used to reach the levels in the lower zone (levels 9 to 33), as in any other building.

Each elevator cabin, whether local or express shuttle, is double. This means that each double cabin - one is immediately placed above the other - serves an even level and an odd level at the same time. This system allows twice as many people to be moved at the same time and requires fewer elevator shafts to accommodate a double capacity of elevators, therefore leaving more net usable area. In a conventional vertical transportation system, more elevators requiring more elevator shafts throughout the building would use almost all the volume available, leaving less lettable space. TAIPEI 101 is one of the few buildings in the world to feature both the sky-lobby system and double-deck elevators for all its elevators. An advanced elevator pre-calling system enhances efficiency.

Three single-deck service elevators and six single-deck car-park elevators complete the vertical transportation system at TAIPEI 101.

19
19A OFFICE SPACE
19B OFFICE SPACE
19E OFFICE SPACE
19F OFFICE SPACE
19C TAIPEI 101 SHOW SUITE
19D OFFICE SPACE

TUNED MASS DAMPER

While TAIPEI 101's structural system met the requirements for reducing building sway and withstanding typhoon wind forces, occupant comfort was still a major issue to be resolved. The solution was a tuned mass damper (TMD) that not only contributes to user comfort, but by incorporating it with the building's architecture and putting it on display with a dedicated observation area, has proved to be a major tourist attraction.

Tuned mass dampers work by stabilising a structure against the motion created by forces of wind, seismic activity, or vibrations caused by the mass movement of occupants of a building or by heavy industrial machinery. The TMD provides a counterweight that moves in one direction as the structure moves in the other, thus reducing the oscillation of the building.

The first use of a tuned mass damper was atop the 59-storey, 279-metre Citicorp Center (now Citigroup Center) completed in New York in 1977 (Hugh Stubbins and Associates, Emery Roth & Sons as associate architect, with structural engineer William LeMessurier and Associates and the Office of James Ruderman).

TAIPEI 101's tuned mass damper is the largest in the world. It has a diameter of 5.5 metres, a gross weight of 660 metric tons and consists of 41 layered steel plates, each 12.5 centimetres thick, welded together on site to form the sphere. The 88th- and 89th-floor TMD observation areas allow the public to closely inspect the workings of the massive gold-painted ball.

The spherical mass block is suspended from the 92nd floor by eight steel cables, 42 metres long and 9 centimetres wide; each cable is composed of more than 2000 smaller cables to ensure flexibility and durability. The cables attain a Security Parameters

Index score of around 9; each can support the weight of the tuned mass damper.

Situated on the 91st floor, the tuning frame supports the suspended steel cables, monitoring building vibration and adjusting cable movement to regularise building movement and optimise damping. The tuned mass damper cuts building vibration by 40 percent, and can on occasion move by as much as 35 centimetres. Moving with the building at the rate of one oscillation every seven seconds, the tuned mass damper's movements are not always visible. Major typhoons (the type that occur once every 100 years) can cause 150 centimetres of movement, mitigated by the specially designed bumper system - eight primary hydraulic viscous dampers situated beneath the tuned mass damper automatically absorb and dissipate vibration impacts.

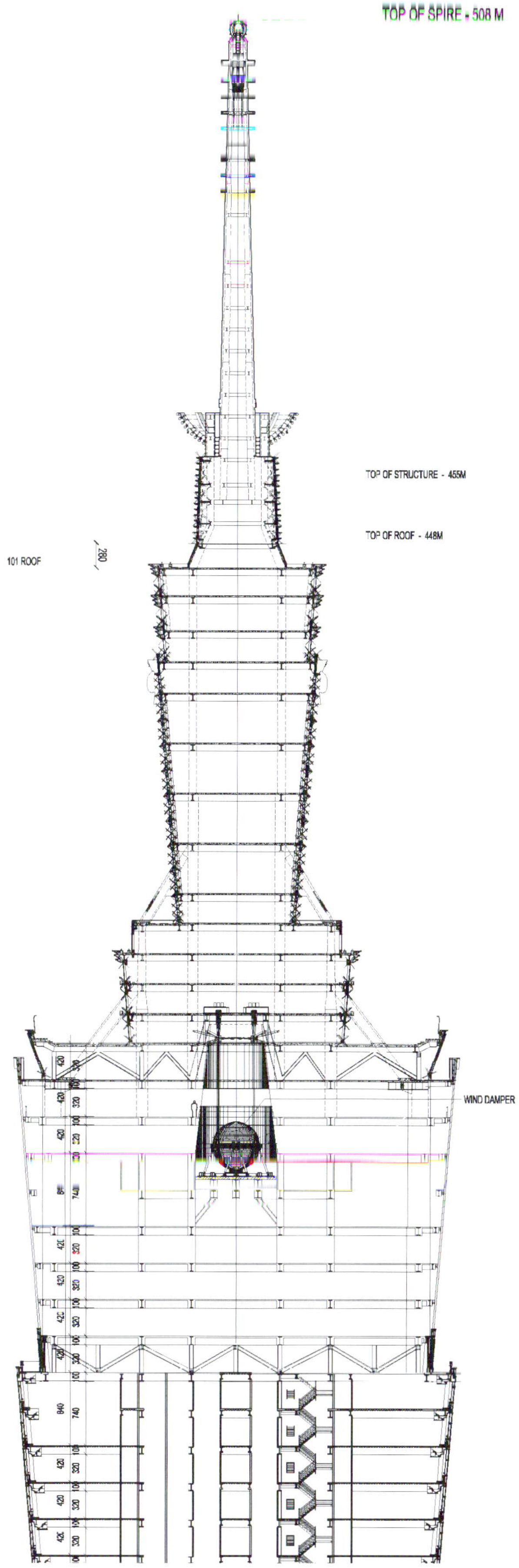

TAIPEI 101's pinnacle, at 450-508 metres in height, features two small tuned mass damper units of 6 tons each. These tuned mass dampers reduce vibration by up to 40 percent, as the pinnacle may oscillate up to 180,000 times a year.

The tuned mass damper was designed, manufactured and constructed by Canadian Motioneering, Inc. at a cost of NT$130 million. Evergreen Consulting Engineering, Inc. was responsible for the overall structural design.

The TMD has gained even more exposure by a joint effort between the TAIPEI 101 Observatory and the Japanese Sanrio Far East Corporation: the creation of 'Damper Baby' merchandise.

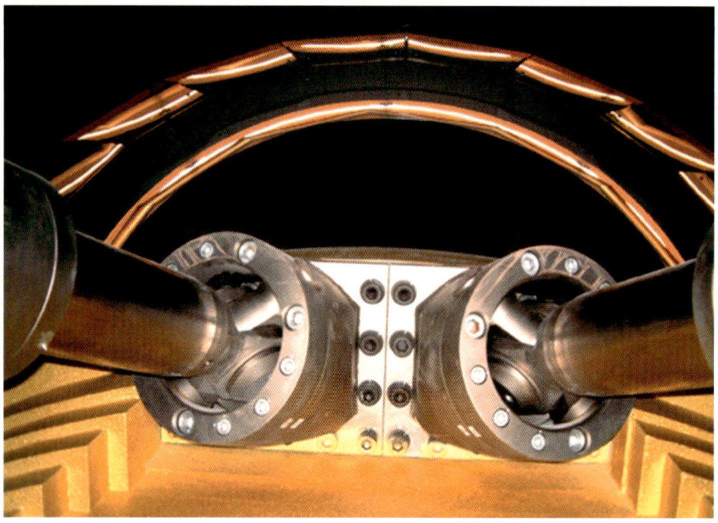

Special effects in the corridor leading to the 88th-floor tuned mass damper observation area.

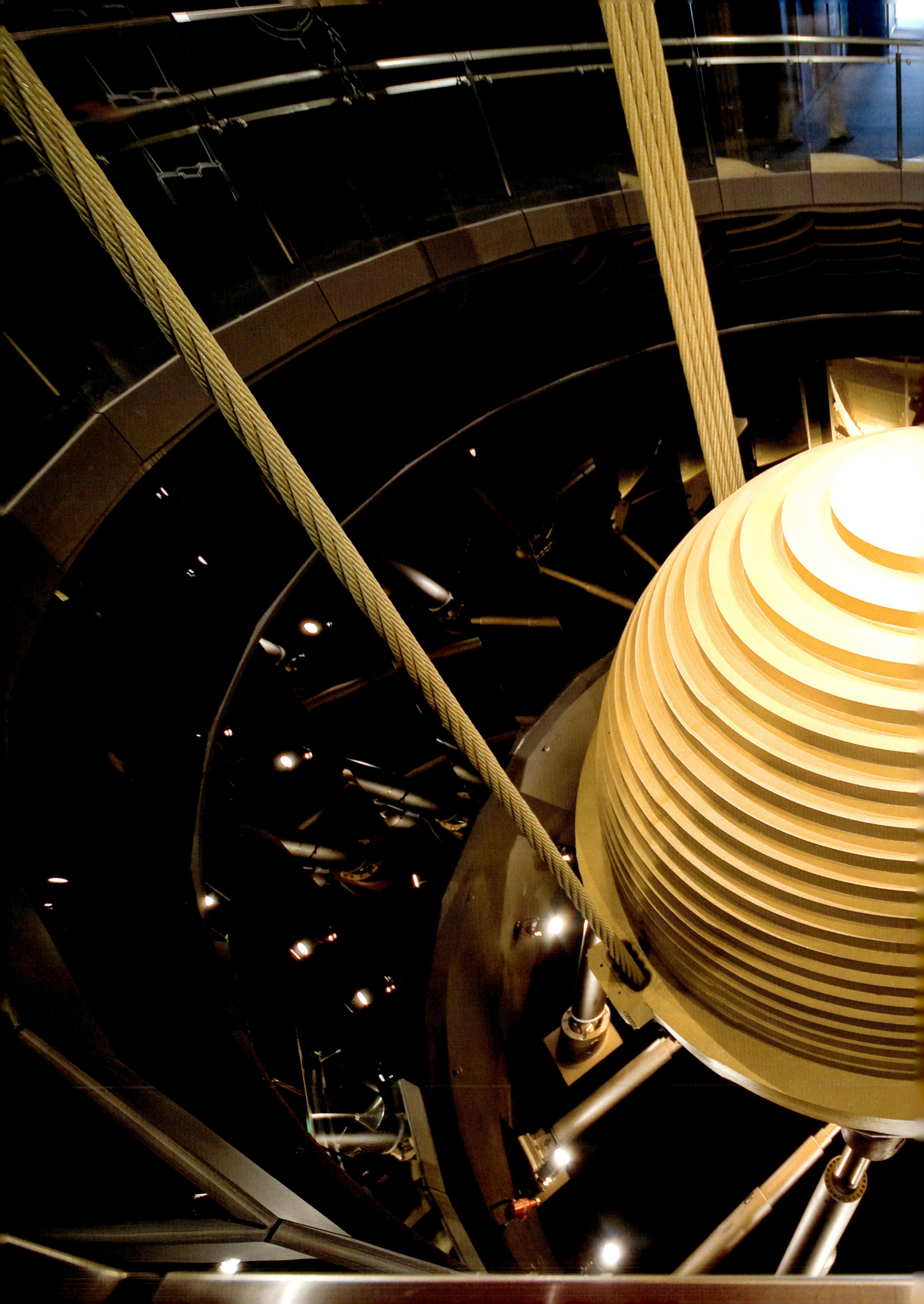

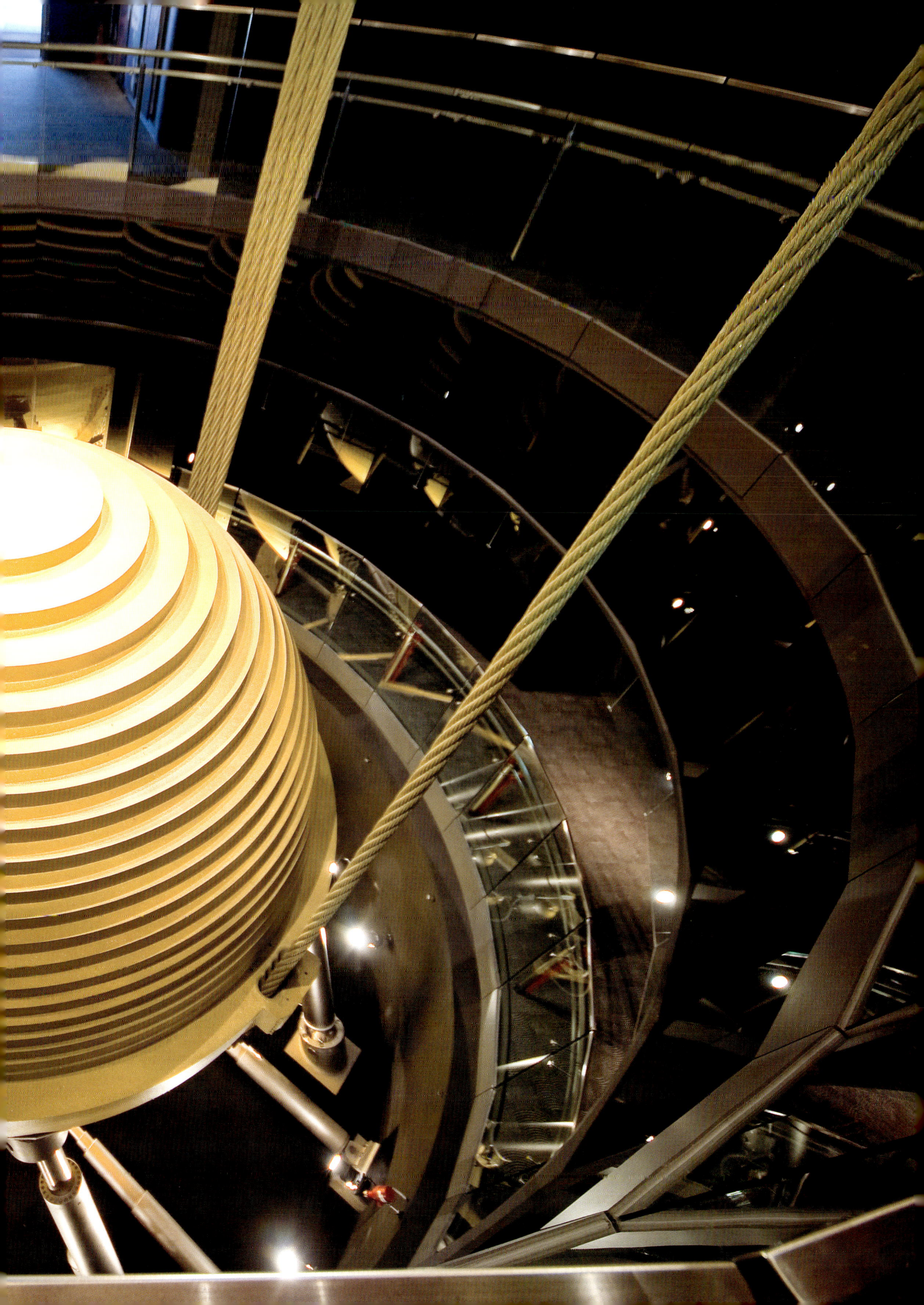

BEHIND THE SCENES AT TAIPEI 101

While the splendour of TAIPEI 101 is immediately obvious, there are a multitude of behind-the-scenes elements that work together to support the tower and its occupants. The Energy Management and Control System (EMCS) monitors and manages all building, fire and life and safety systems. In addition to occupier safety, minimising the impact on the environment is a primary objective.

Power supply, water supply, ventilation, air conditioning, lighting, security systems, telecommunications and other building management systems are all monitored inconspicuously, as are hazards such as unauthorised access, power failure, fire and earthquake activity. In the highly sophisticated building management control room, data is collected from thousands of collection points around the building, ensuring immediate response to potential problems.

Building management central control room.

Panel in the central control room.

火警監控系統(FAS)
預備迴路
放水槍監控系統

Fire control room.

Fire hydrant system pump room.

Fire protection system

TAIPEI 101 has one of the most comprehensive fire systems ever implemented, ensuring a safe and secure environment. Firewalls, shutters and protective wall fillings divide the building into discrete fire zones. Non-flammable materials are used for the base building and all fit-out work. Detectors located throughout the tower continuously take air samples while the automatic sprinkler system is linked to the water tanks located in the mechanical plant rooms. The sprinkler system is gravity-fed so that a power failure cannot interrupt the water supply. In the event of a fire, integrated hazard management and safety systems are automatically activated, fire protection doors closed, smoke extraction systems switched on and escape routes ventilated.

Each office floor has two pressurised fire safety corridors directly connected to separate pressurised emergency staircases, which provide access to the two refuge rooms located on each mechanical floor located every eight levels. Mechanical floors above the 34th floor have an outdoor refuge balcony. All the office floors and key escape routes are protected by smoke extractors.

Pressurised corridors and staircases.

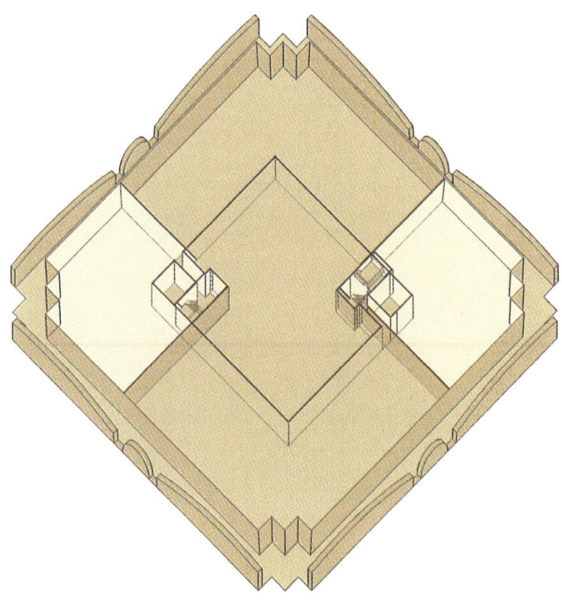

Mechanical floor refuge areas and balcony.

Security systems

TAIPEI 101 is protected throughout by a full team of professionally trained security staff on site 24 hours a day, 365 days per year. The building boasts one of the world's most advanced, customised, integrated electronic security systems, designed to provide continuous control and monitoring of the tower and to give tenants and visitors complete peace of mind. The surveillance system, which includes more than 600 CCTV cameras and 400 card readers at doors and elevators, transmits real-time security data of all main entrances, elevators and public/common areas through an optical cable network to the central control centre.

Managing visitor information, the Visitor Access Kiosks (VAKS) integrate a visitor management server, a visitor card dispatching machine, a visitor imaging system, optical turnstiles and computers operated by security personnel. The system keeps a comprehensive record of the identity and destination floors of all visitors. Visitors are required to call the tenant they wish to visit from a kiosk located in the lobby. If the tenant authorises the visitor then a card that gives access to the turnstiles is issued, cutting administration time, preventing unsolicited visitors and improving security.

The elevators and entrances to TAIPEI 101 are electronically controlled by a smartcard access system that operates 24 hours a day. Occupants and visitors without access cards must check with the personnel at the information counter at Level 1 to confirm their identity; only then will they be allowed to gain access to the tower.

訪客發卡機
Visitor Access

Telecommunications

TAIPEI 101's telecommunications network is managed by Chunghwa Telecom. This advanced, integrated telecommunications infrastructure includes two fibre-optic backbones, from separate Chunghwa Telecom central stations, which enter the building at different points. Internal distribution from the main distribution frames (MDFs) located on the 17th and 66th floors is via separate 10 Gbps backbones to an intermediate distribution frame (IDF) located on each office floor.

TAIPEI 101 is equipped with three alternative back-up means of communication: microwave, free space optics and satellite (to two earth-stations). The entire building is mobile accessible, including elevators and basements. The public areas provide Wi-Fi access.

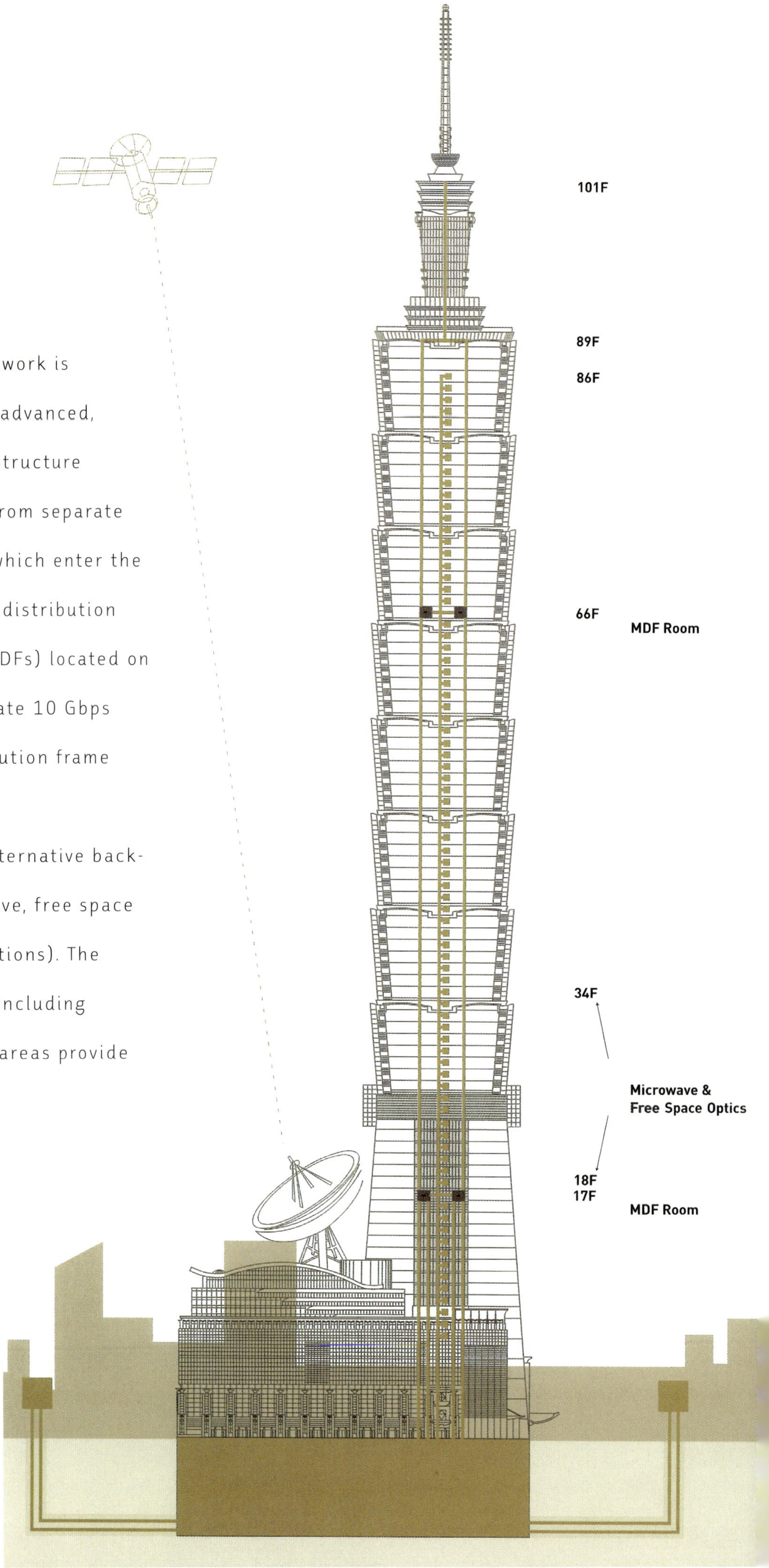

Telecommunications plan.

Power supply system

TAIPEI 101's integrated power management system ensures efficient power consumption. Within the tower, separate dual risers from two Tai Power substations, plus three transformer rooms, each with two transformers, deliver power to the mechanical floors. Each office floor is connected to two mechanical floors, minimising the risk of connection failure.

In the event of a power failure, the building management system automatically starts the emergency generators, which provide about 70 percent of the tower tenants' power consumption needs. An Uninterruptible Power Supply (UPS) ensures that the building's control and security systems remain in operation before the emergency generators start.

HVAC system

TAIPEI 101's ventilation and air conditioning is designed for optimum occupant comfort and energy efficiency. The HVAC system consists of centrifugal chillers, the glycol chiller and the ice storage tank. The chiller produces chilled water for air conditioning use. The glycol chiller utilises off-peak power at night to produce ice to be stored in the ice storage tank that is then used to supply chilled water during the day to make up insufficient chilled water volume from the chiller. The system includes 40 to 52 Variable Air Volume (VAV) boxes on each floor, for flexible and adjustable air volume to ensure occupant comfort. The air supply and return system is integrated with the building's lighting system to ensure better distribution of air. VAV boxes helps make air conditioning systems more efficient by regulating the amount of cooling targeted towards any specific room or area; a room whose windows face the sun for most of the day, for example, will have a higher cooling load than a room in the shade.

Generator room.

VERTICAL CITY

TAIPEI 101 is the centrepiece of the Xin Yi mixed-use area. It is also a living vertical city. People fortunate enough to work in or visit this iconic building can find everything under one roof, from luxury fashion stores to gourmet dining; from financial services to a destination-type bookstore mixed with a city square. On the lower level of the Mall, the Grand Market store caters to all daily needs while a series of food and beverage outlets complement the high-end restaurants available on the fourth level. Office workers and visitors can soak up the extraordinary atmosphere and architectural aesthetics while shopping and relaxing in pure comfort. The TAIPEI 101 Mall joins the ranks of world-class shopping districts with flagship stores and new design concepts.

The 77,000-square-metre Mall, located within the podium, contains 163 stores with a mix of premium brands, a range of restaurants and cafés and the largest food court in Taipei. The Mall is accessible from the basement, the ground level, the second level via covered pedestrian bridges, and tower occupants can access the Mall from levels four and five.

B1 contains the 'Grand Market', featuring Jason's Market Place, a food court that seats 1200 and offers an amazing array of international foods, plus leisure and convenience stores.

1F is 'Avenue 101', featuring international popular brand fashion, accessories and cosmetics. Intimate cafés provide a relaxing shopping environment.

2F is the 'Designers' Walk', featuring international designer brands and the Sogo 101 department store. It is connected by pedestrian skybridges to the World Trade Center, the Grand Hyatt Taipei and the New York New York shopping centre.

3F is the elegant 'Galleries Couture', featuring couture brand flagship stores offering the most prestigious luxury brands in Taiwan. It brings Fifth Avenue glamour and Paris catwalk fashion to Taipei.

4F is the exquisite 'City Square', more than 1670 square metres in area and 40 metres high. Featuring natural lighting and arboreal canopies, it is Taipei's most striking space for music, art, fashion and entertainment. It also includes restaurants, cafés, and international bookseller, Page One.

5F includes several banks offering personal banking and financial services, luxury electronics retailers and beautiful boutiques offering unique Taiwanese products.

NINE WEST
JUST ARRIVED
GODIVA

In the tower section, a convenience store, florist and the café corner are conveniently located on level 35. A state-of-the-art conference centre located on level 36 provides a meeting place with unmatched views over the city. The restaurants and bars located on levels 85 and 86, as well as the observatory on levels 88, 89 and 91 create a leisure ensemble where the magical views add to the high-end service provided. The 91st-level outdoor observatory offers an even more dramatic experience.

If life is bustling from within, the excitement can also be experienced outside TAIPEI 101. In the evening, the lighting system of TAIPEI indicates the day of the week and has transformed the giant beacon into a personal diary assistant for all Taipei citizens. On special occasions such as New Year's Eve or Valentine's Day, special illuminations and special effects transform TAIPEI 101 into the city's ambassador to the world.

The TAIPEI 101 brand

Diana Chen, Chairperson of TAIPEI 101 especially cares when it comes to identity and branding of the building. By its very presence, whether showcasing an amazing light show or standing serene, it sends its greetings as a friendly neighbour to Taipei and the rest of the world.

While Ms Chen reminds us of the bi-cultural Oriental aspects behind the design of the project or the Western technology used to achieve such an endeavour, she explains the importance of branding as another major aspect of the project. If TAIPEI 101

Ms Diana Chen, Chairperson TAIPEI 101.

through its own identity and services provided is becoming a brand on its own, the aim, through co-branding, is to use the TAIPEI 101 name to promote other brands, in turn promoting the TAIPEI 101 name throughout the world. When French luxury fashion brand Celine celebrated its 60th anniversary

with a fashion show at TAIPEI 101, or when Versace, like Celine, used the building to project its illuminated name along the building's huge façades, it not only gave visibility to these international brands but it also brought the TAIPEI 101 brand to a larger international audience. These types of activity 'introduce TAIPEI 101 globally', says Diana Chen. The relationship with these different brands gives more style and personality to TAIPEI 101, 'like a human being', she adds. When describing the parallel between the New Year's Eve countdowns in Times Square, New York and at TAIPEI 101, Ms Chen enthusiastically explains how massive crowds of people gathered at the foot of TAIPEI 101, which has rapidly become Taipei's - and even Taiwan's - modern symbol. It is part of 'common memory', concludes Diana Chen, and is well and truly established as the symbol of Taipei throughout the world.

Cartier
Cartier
PAGE ONE
PAGE ONE

Dior
Dior
Dior

Dior
COACH
COACH
GOLAY
OMEGA

BVLGARI

ony style
ony style
BRAVIA
OMEGA
Hour Vision.

LIGHTING TAIPEI 101

The specialised lighting employed throughout TAIPEI 101 creates a dazzling array of memorable effects. A centralised lighting system operates lighting in the podium, the main lobby of the tower and the tower exterior, from the 25th floor to the top. The system has built-in lightning protection devices to prevent high-altitude short circuit problems and pre-programmable time clock functions that enable automatic lighting resetting in case of a power failure.

At night the soft glow cast by TAIPEI 101 illuminates the sky above Taipei. Sections of the building glitter like facets of a diamond, with different levels reflecting light at various angles to present a variety of highlights.

Since the tower opened, it has been a colourful addition to the Taipei skyline. The seven colours of the rainbow were chosen to light up the building at night,

2008

with each colour representing a day of the week: Monday - red; Tuesday - orange; Wednesday - yellow; Thursday - green; Friday - blue; Saturday - indigo; Sunday - purple.

TAIPEI 101's lights are also used in creative ways, to send messages to the residents of Taipei and beyond and to promote various events. Among the regular events is Mother's Day, which is celebrated by lighting up the north- and west-facing exterior with 'Mom' in Chinese and English characters during the week leading up to the day. The tower supported the breast cancer awareness pink ribbon campaign in a joint venture with Estée Lauder and showed its support for the Taiwan Society of Wilderness 'Lights out for Summer' campaign by turning off its lights for a period of time.

2008
TAIWAN

2008
2008

The event of the year, however, is New Year's Eve. In a short period of time, TAIPEI 101 has become the place in Taipei to celebrate. After the countdown, colourful fireworks are launched from the outdoor section on every eighth floor, to wish everybody a happy new year.

Golden flashes are released from the pinnacle.

Flower clusters are
shot upwards,
painting the sky
with brilliant light.

The display ends with computer-controlled silver shots that begin at the bottom of the building and end at the top, as if an awakened dragon had wrapped itself around TAIPEI 101 for the new year.

PUBLIC ARTWORKS

With more than 0.5 percent of the total construction budget set aside for public art, TAIPEI 101 has created a record public art project fund in Taiwan. Artists from eight countries - the United States, Canada, Britain, Germany, France, Singapore, Taiwan and Hong Kong participated in the project, the aim of which was to provide a public art gallery accessible to all visitors to TAIPEI 101.

LOVE by Robert Indiana, USA, 1928-
Aluminium

Prominent among the artworks is the big red *LOVE* sculpture by American Pop Art artist Robert Indiana. Located on the plaza, near one of the tower's entrances, *LOVE* is probably one of the world's most famous contemporary pieces of art created since the beginning of the series 40 years ago.

_1-0_ by Robert Indiana, USA, 1928-
Aluminium 45 cm x 46 cm x 25 cm (10 pieces), 2002
Robert Indiana's sculpture in the lobby of TAIPEI 101 uses multicoloured numbers to suggest the conduct of world trade and the circle of human life.

Dragon Feature - The Finale Show **by Crystal Fountain Inc., Canada**

Another popular artwork is the Dragon feature fountain located on the plaza facing one of the main entrances to the shopping mall. At the centre of the 17.5-metre-long, 7.5-metre-wide fountain is a light globe, which works in conjunction with multiple spouts to create the image of a pearl. The fountain tells the story of two dragons with a pearl, which is a sign of prosperity in Chinese culture. Modulated by computer, 200 nozzles create the dynamics of soaring dragons. The pearl is fashioned from 25 large nozzles and 78 colour-changing LEDs. Thirty computer-controlled patterns mix and match to produce numerous variations. Patterns on the rocks radiate outwards in contrast with the outer surface treatment. In the summer, soft music can be heard around the fountain and it is fitted with a safe mode to enable children and parents to enjoy it during the no-show hours. At regular show hours, the spouts modulate their output to the rhythm of the music, which is played through the Bose audio system.

TAIPEI 101

Seal by Cheung Yee, Hong Kong, 1936-
Bronze 330 cm x 330 cm, 2002

For visitors entering the office tower lobby, it is impossible to miss the large 3.3- by 3.3-metre *Seal* designed by Hong Kong artist Cheung Yee. The seal, inlaid in the floor, is inscribed with layers of pious blessing characters: 'promotion, progress, abundance, well being, modesty, persistence and learning from the moves of nature'. In the centre of the matrix is 'eternal treasure'. This seal brings good wishes to the users of the building and assures the integrity of the businesses within.

***City Composition* by Jill Watson, Britain, 1957-**
Bronze 170 cm x 120 cm x 90 cm (3 pieces) 2002

British sculptor Jill Watson's sharp powers of observation have captured the expressions of the residents of Taipei in these sculptures of three family members. Her sculptures are a reminder to 'slow down and enjoy the scenery'.

TAIPEI
101

Global Circle **by Chung Pu, Taiwan, 1947-**
Black granite, white marble outer circle; diameter 2140 cm, 2002

The black granite ring is embedded in the Social Plaza of the shopping mall; white inscriptions of names of cities, from celebrated metropolises to unknown towns, are sprinkled along its border, symbolising the place of Taipei in the context of globalisation. Sparkling in the sunlight, the *Global Circle* echoes the constellations of the cosmos.

Model of City Square.

***Dialogue Between Yin and Yang* by Rebecca Horn, Germany, 1944-**
Stainless steel mirror, binocular, iron
Base: 450 cm x 450 cm x 15 cm;
Cylinders on base: 150 cm~220 cm, 2002

The ancient Chinese concept of Yin and Yang is dynamically represented through Rebecca Horn's delicately designed mechanism. Mirrors and binoculars are metaphors for Yin and Yang respectively and through their focusing, reflecting and imaging, the work evokes meaningful dialogue within itself, and with its spectators and the environment.

Between Earth and Sky by Ariel Moscovici, France, 1956-
Rose de la claret granite, 100 cm x 130 cm (6 pieces), 2002

In this arrangement of street furniture, Moscovici creates a galaxy of human warmth. Consisting of eight hemispheres that symbolise reincarnation and eternal life, the sculpture invites passersby to take a rest or have fun. It has proven to be a popular meeting point in the city.

The Rising Sun – Shooting Stars **by Crystal Fountain Inc., Canada**

This waterscape occupies the gentle slope on Sung-Ren Road. A 90-centimetre-diameter granite ball at the top rolls as the water whirls and flows around the angles created by the leaning egg-shaped fountain. Around the egg, water jets light up the night. The three-dimensional granite assemblage is of human height and proportion, designed to create a congenial atmosphere that puts pedestrians at ease. A wind-speed sensor reduces the jet power when the wind is particularly strong.

UNITED COLORS
OF BENETTON.

Partners Monument **by Florence Ng**

As a tribute to all the people involved with the TAIPEI 101 project, including investors, designers, consultants, owner's team and workers, their names were inscribed on colourful glass bricks, which were then assembled as a tangram to create the *Partners Monument*. Designed by world-renowned glass artist Florence Ng, the monument is installed on the Xin Yi Road side of TAIPEI 101, facing the main entrance. The seven colourful, bright glass pillars become light boxes that glitter at night.

A SERENE PRESENCE

TAIPEI 101 is a tranquil backdrop to life in Taipei. The familiar silhouette is visible on the horizon from many miles away; up close, its benign presence hovers and protects the residents of Taipei as they go about their daily tasks.

63828
741

2008
TAIWAN
千米水岸奇蹟
日勝生集團 小碧潭捷運站 8911-3333
首泰大直
CITY VOICE
明水路606巷 8502-1000
冠德
遠見
150坪一層一戶

安 和 路
一 段 | 二 段
AnHe Rd.
Sec.1 | Sec.2
信義路四段
XinYi Rd. Sec.4
公車專用道
照相
相友店
HILLTOP

THROUGH THE LENS

TAIPEI 101 provides photographers, both amateur and professional, unsurpassed opportunities for spectacular photography.

Its continually changing appearance, due to the time of day, season and weather, means that no two photos are ever exactly the same. The scope for photographic experimentation and artistic expression is boundless.

DESIGN STUDIES

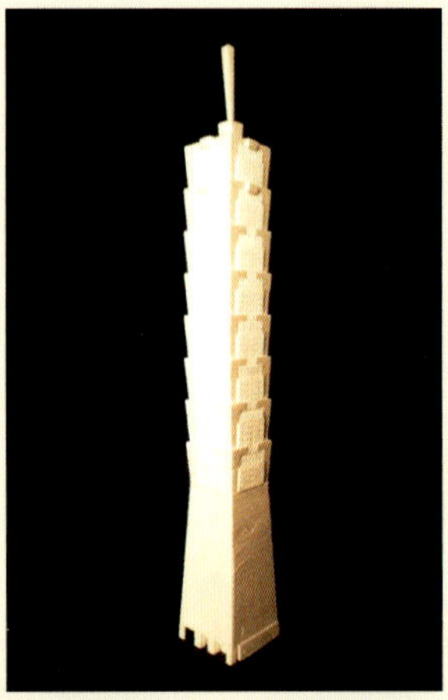

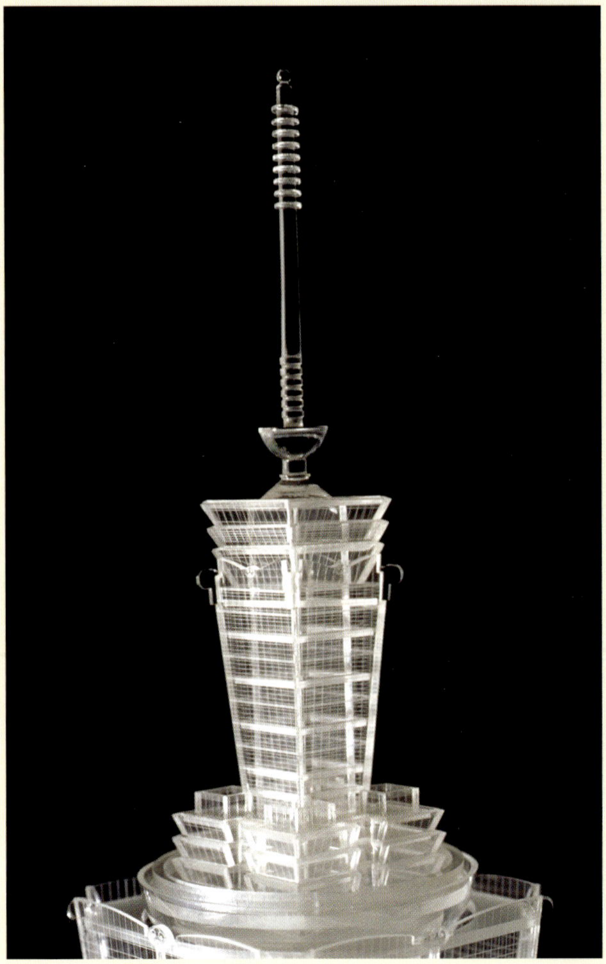

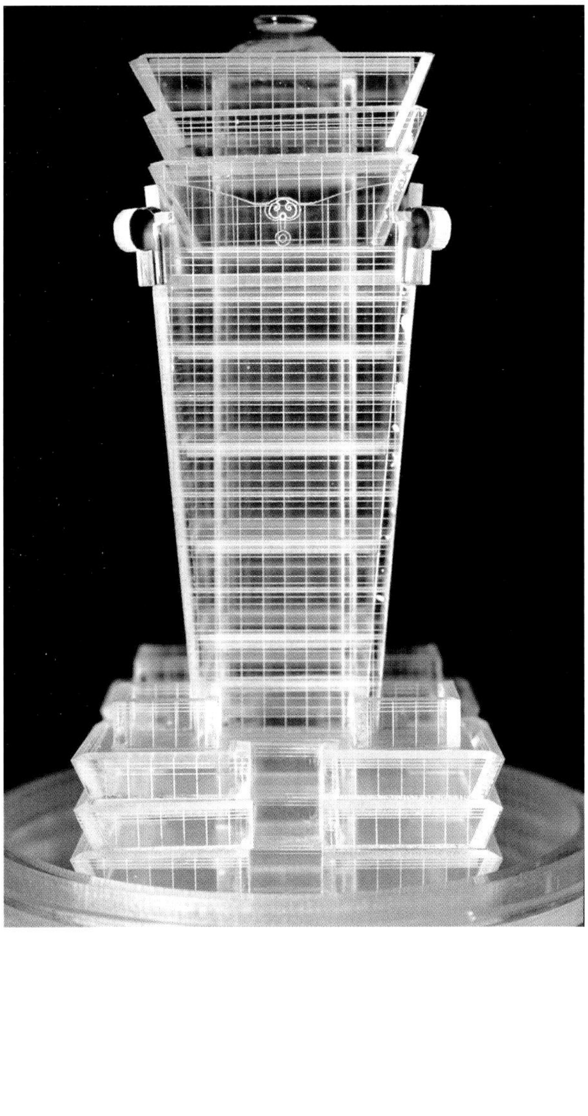

This model included a removable roof, so that the interior detail of the shopping mall could be easily viewed.

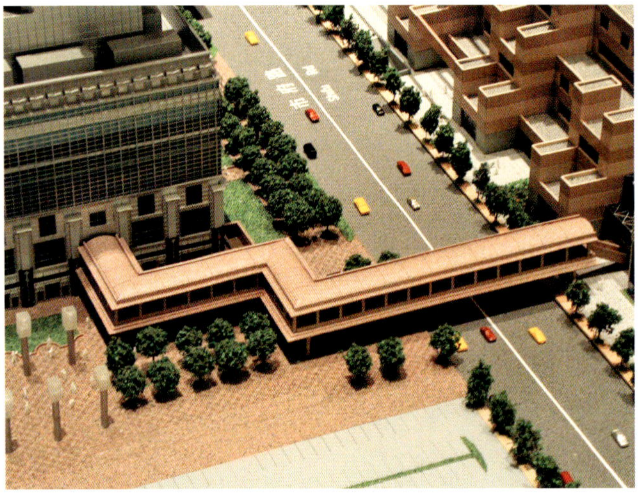

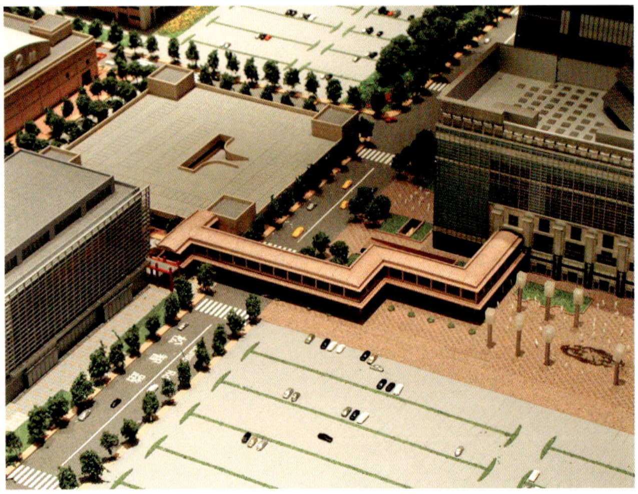

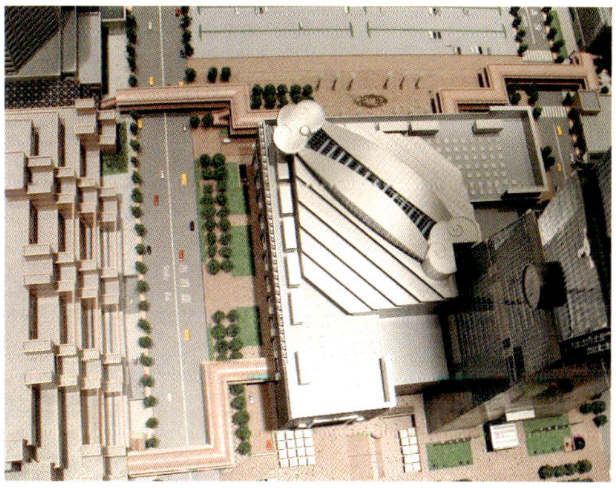

FLOOR PLANS

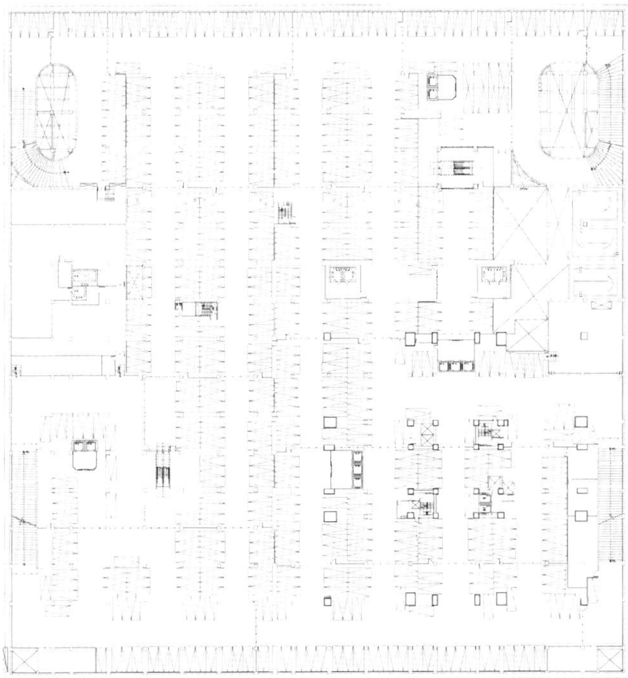

Basement level 3

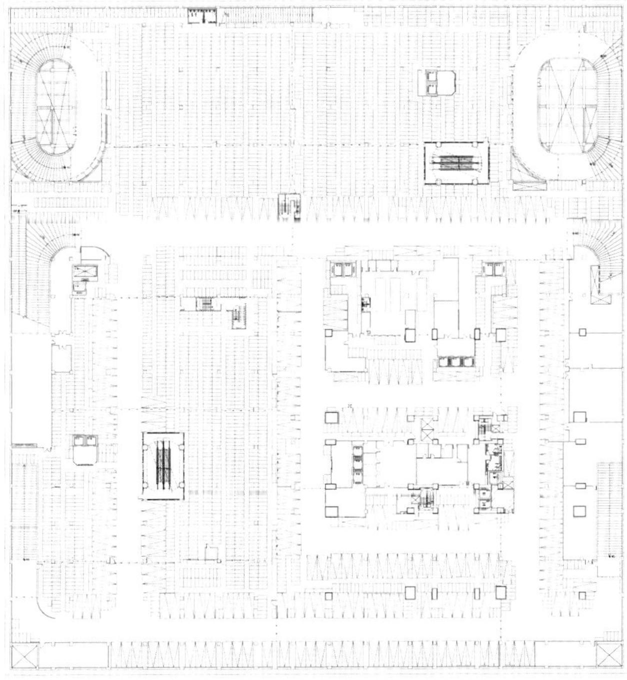

Basement level 2

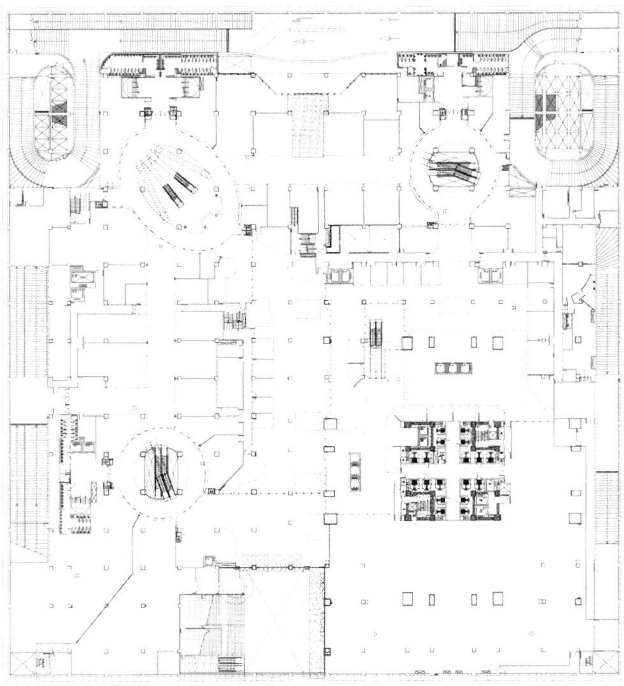

Basement level 1

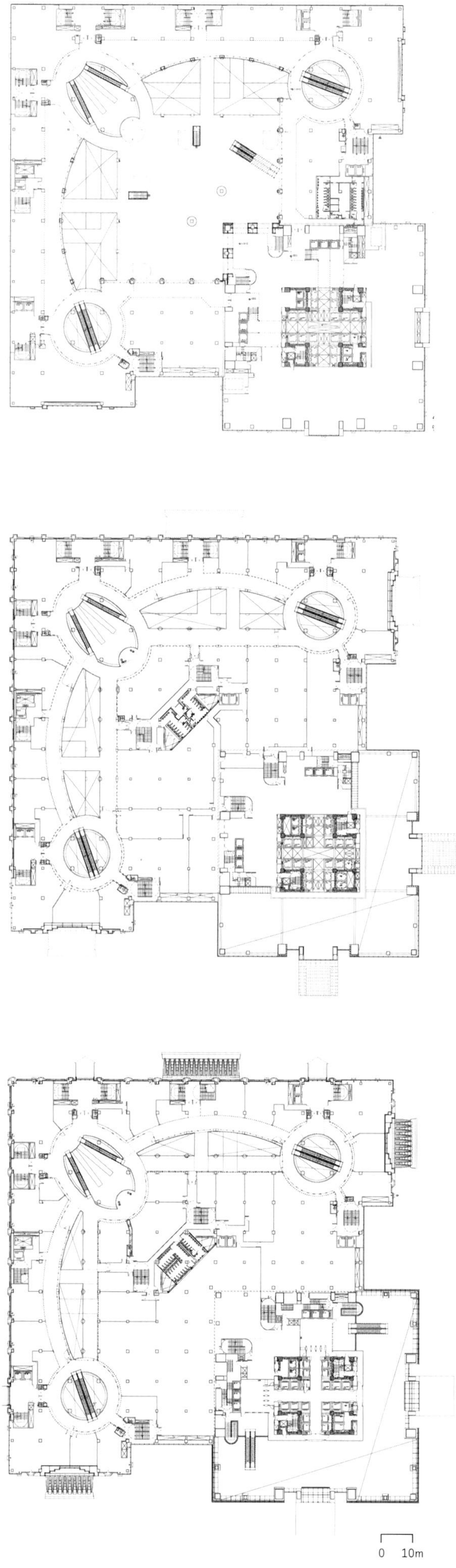

Floors 2 (bottom), 3 (middle) and 4 (top): tower and shopping mall.

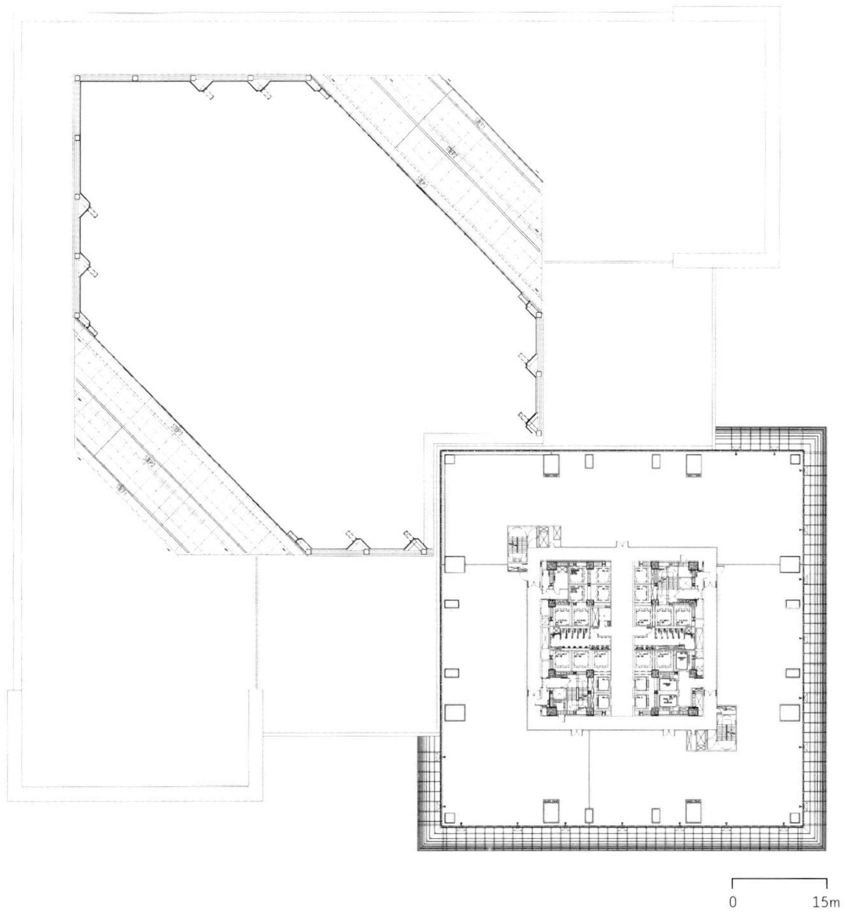

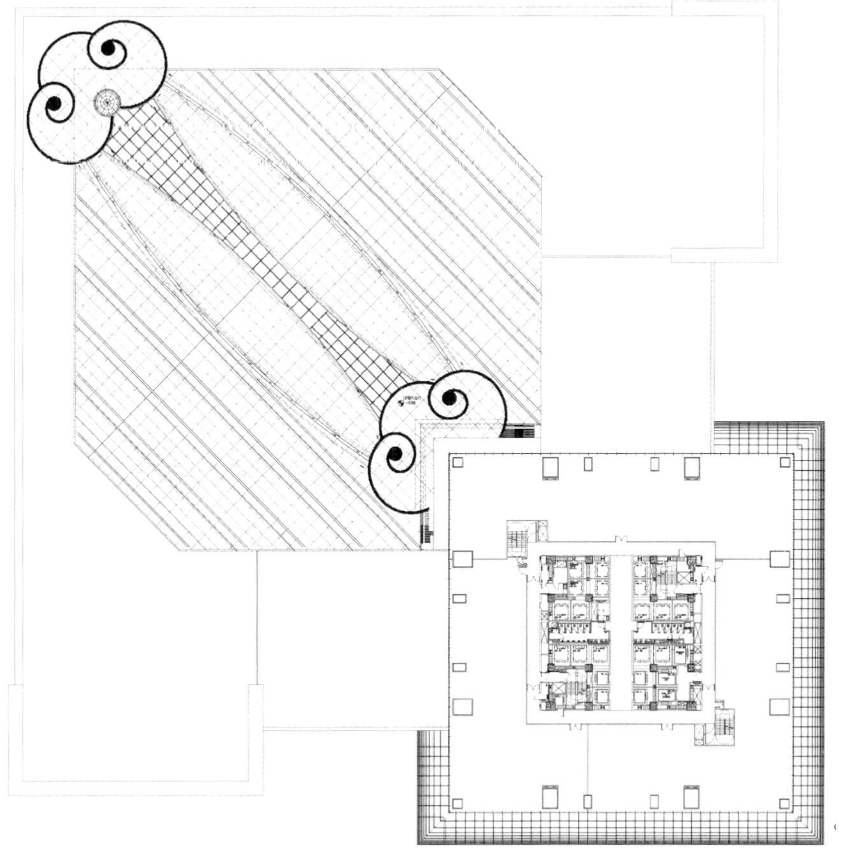

Floors 13 (bottom) and 14 (top): podium and tower low zone.
Typical low-zone floorplate is 2810-4300 square metres.

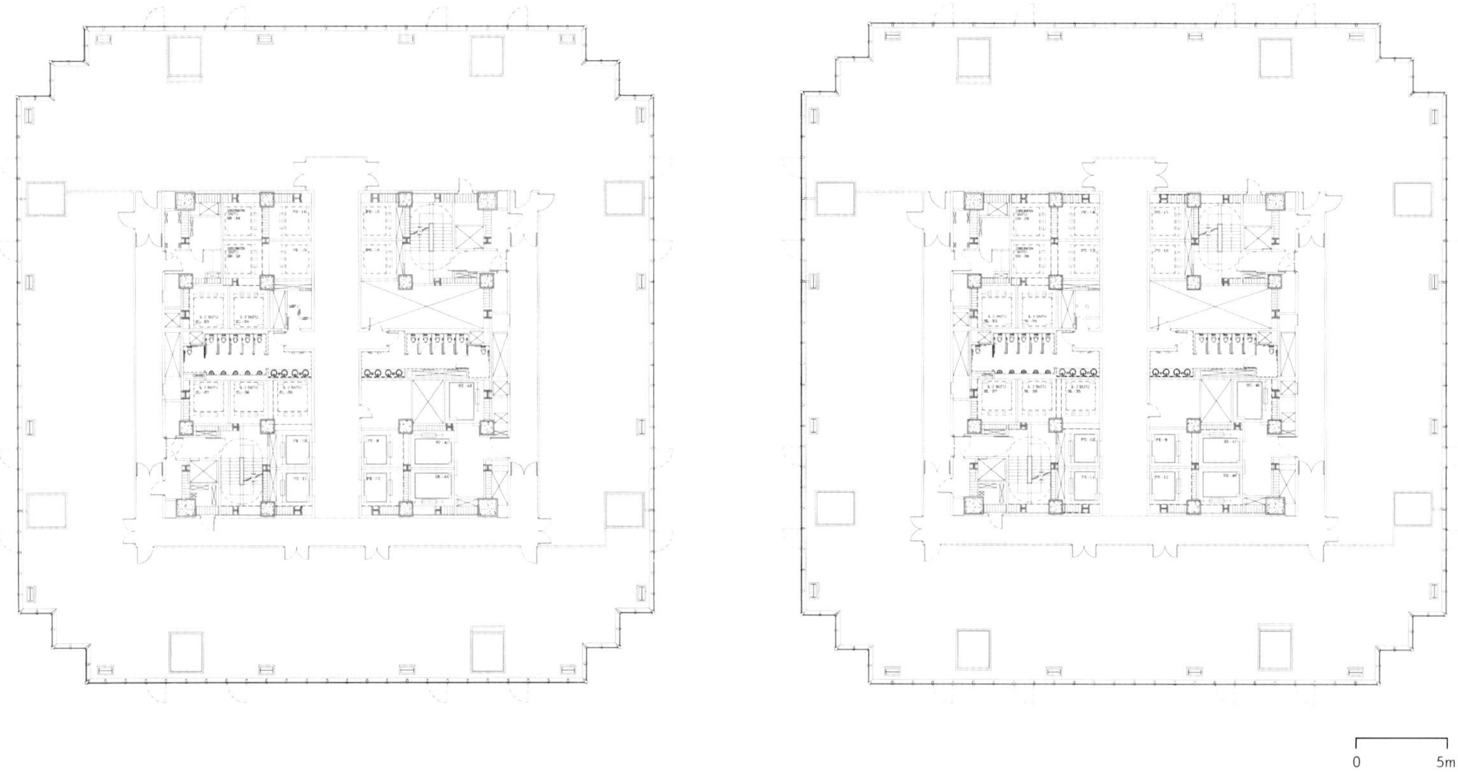

Floors 42 (left) and 43 (right): typical mid-zone tower floor plans. Typical mid-zone floorplate is 2500-3305 square metres.

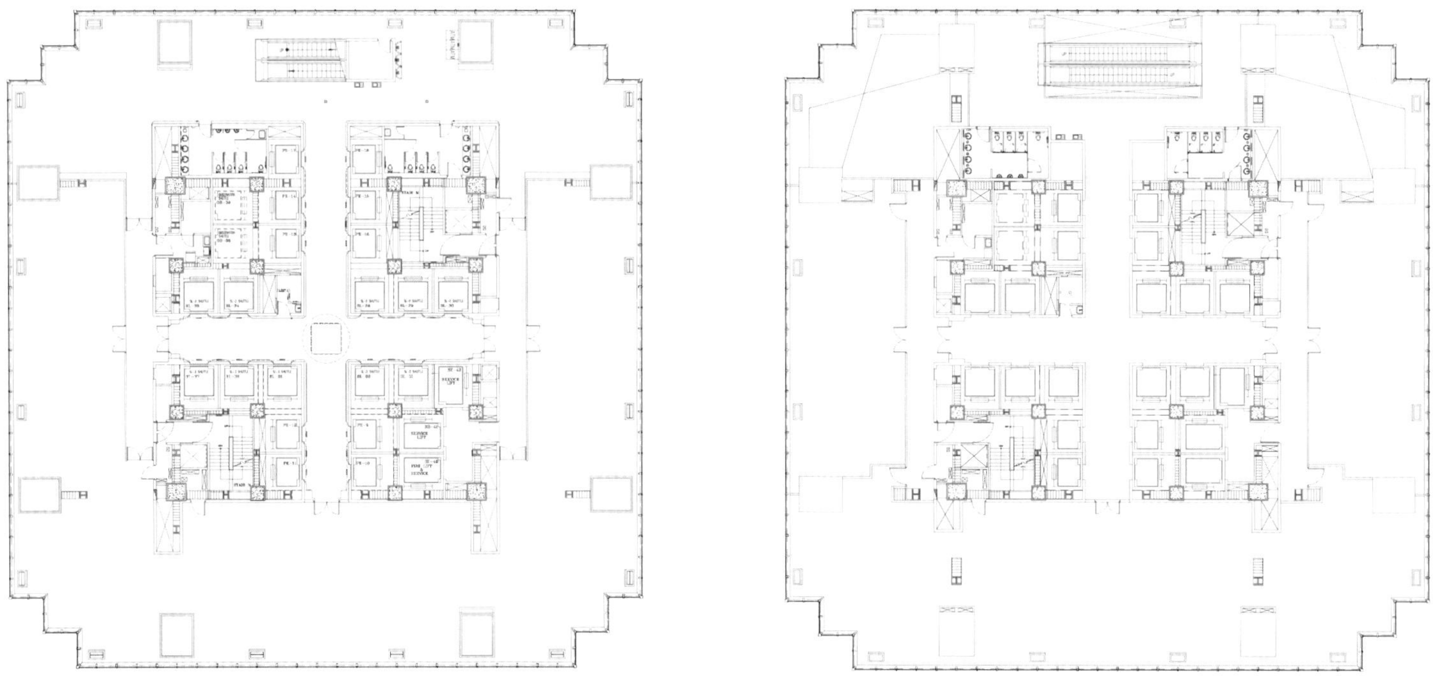

Floors 35 (left) and 36 (right): sky lobbies; floor 36 houses the International Conference Center.

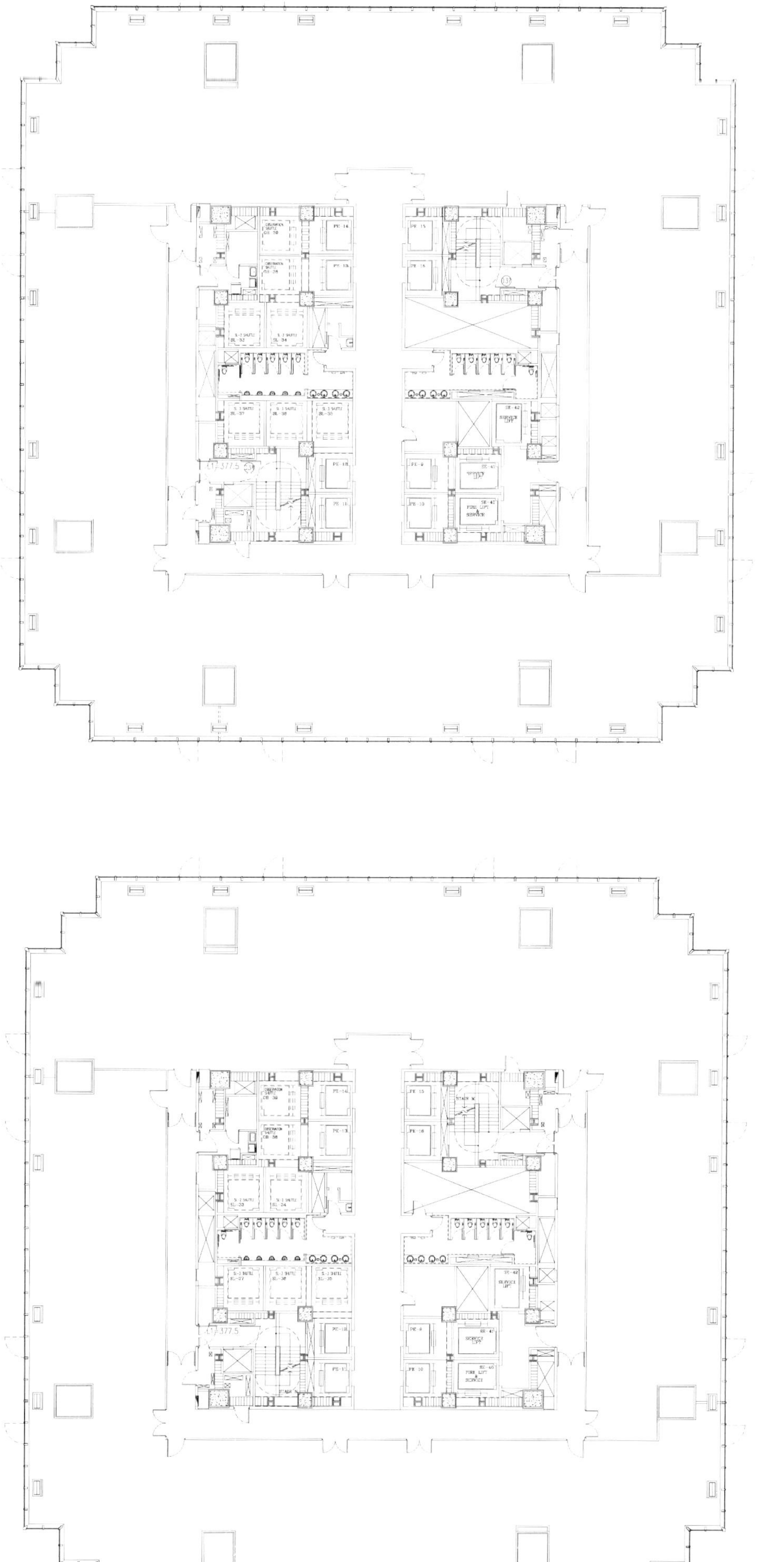

Floors 47 (bottom) and 48 (top): transfer floors, mid zone.

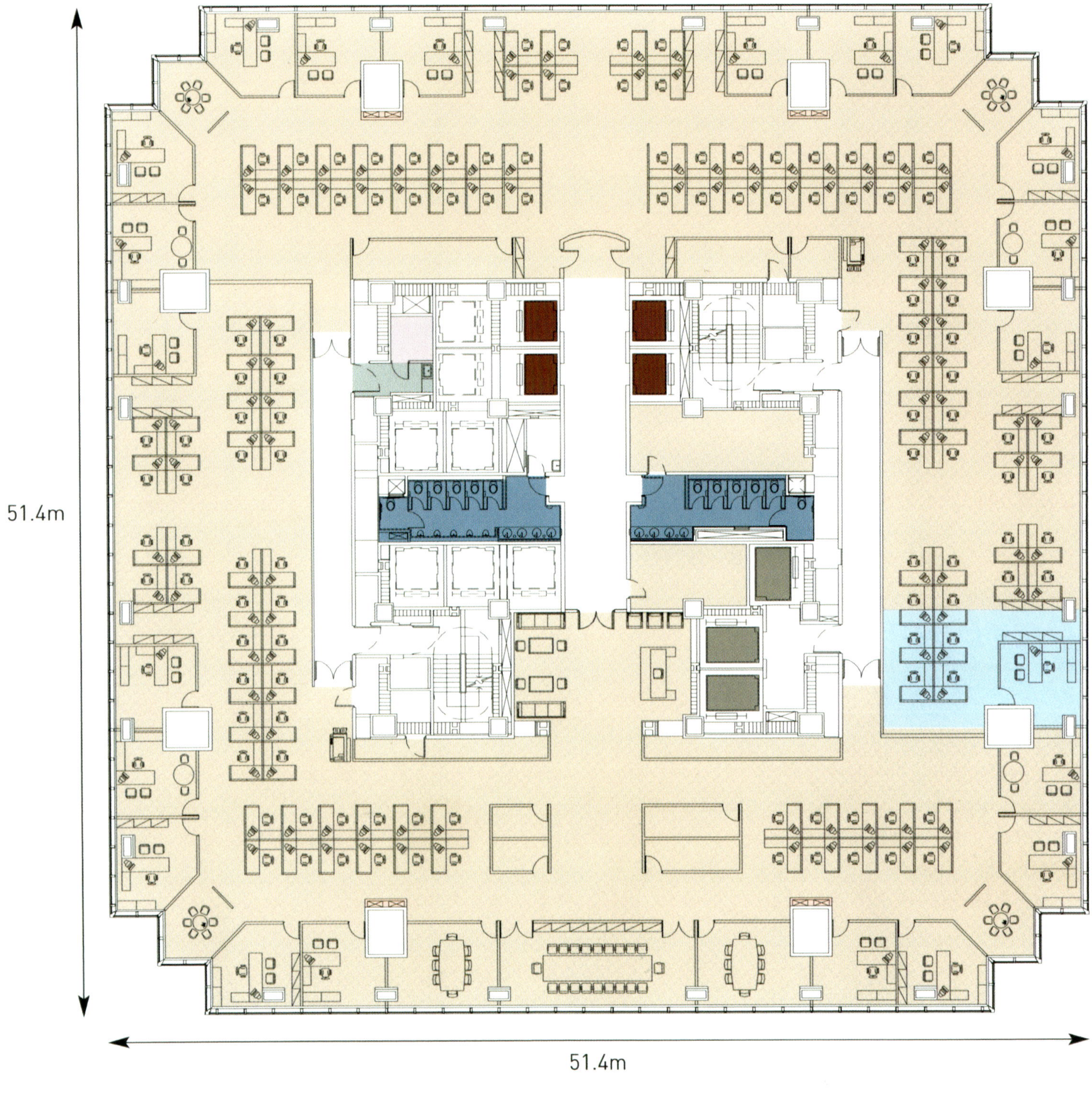

Typical office fit-out plan.

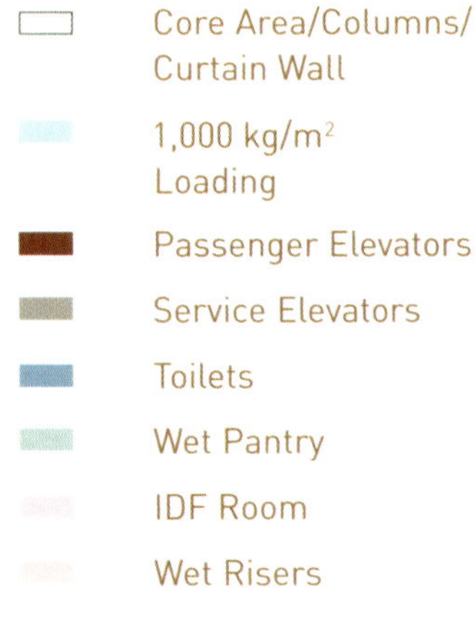

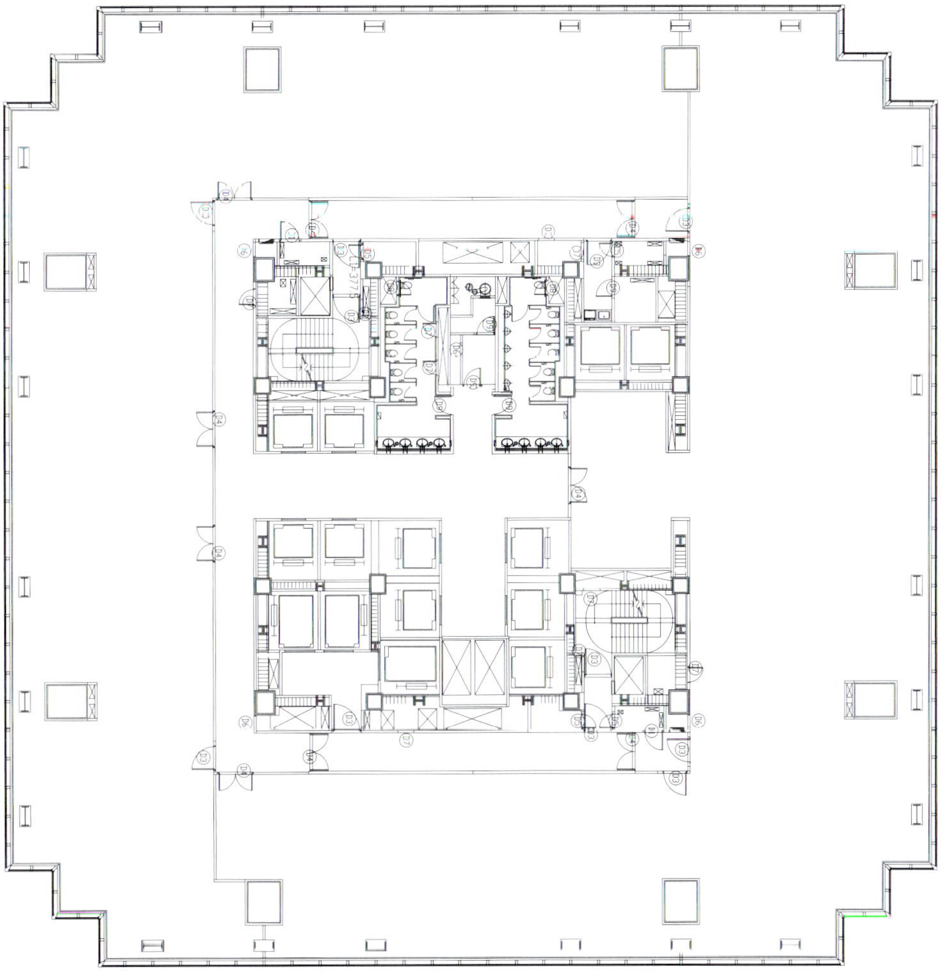

Floor 71: transfer floor for high zone.

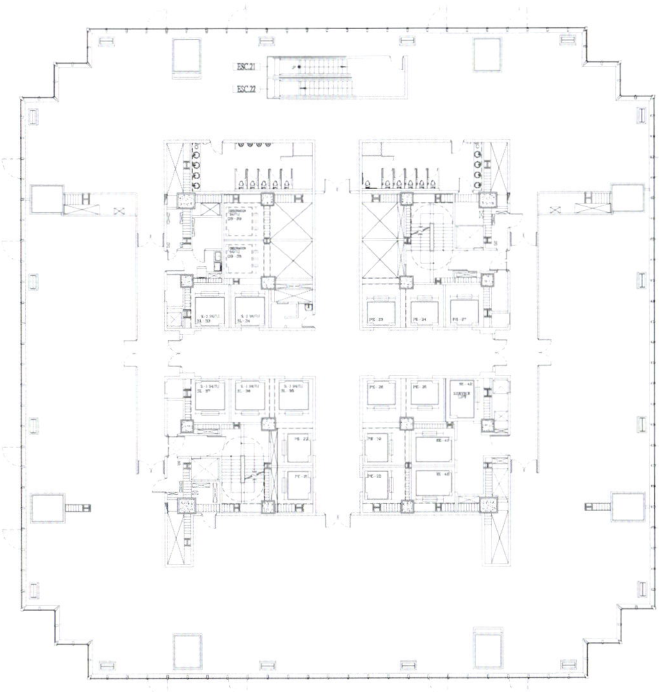

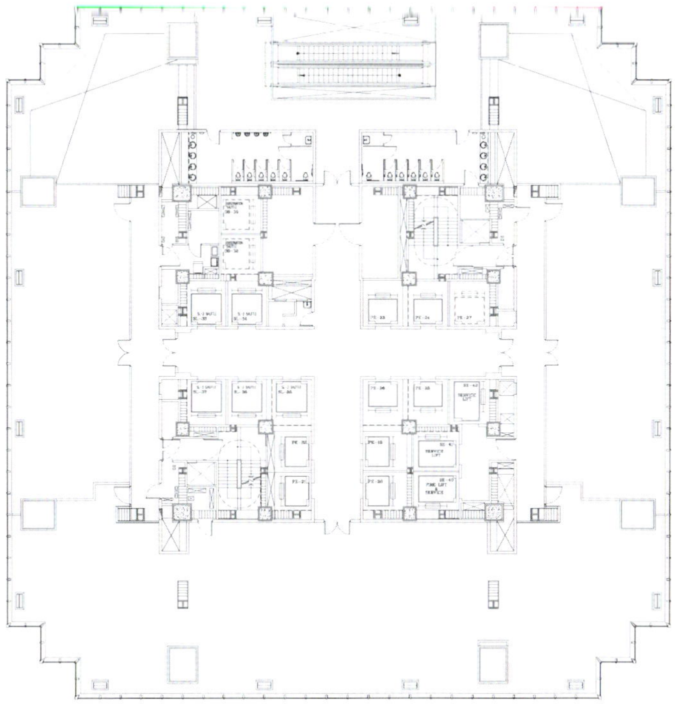

Floors 59 (left) and 60 (right): sky lobbies, high zone.
Typical high-zone floorplate is 2645-3570 square metres.

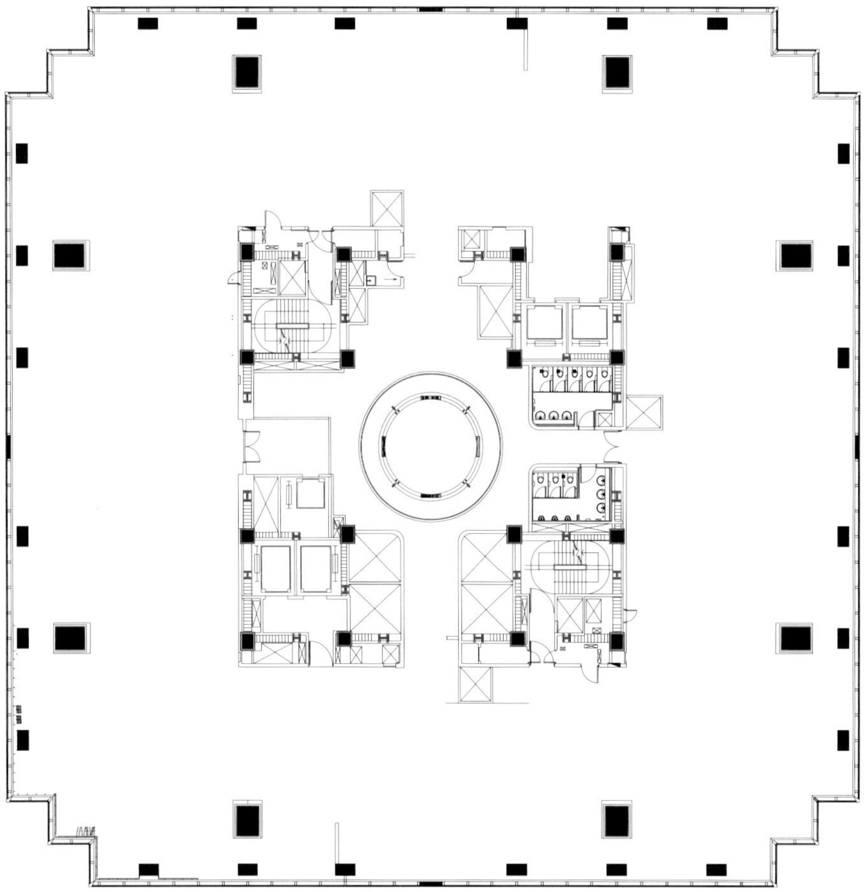

0 5m

Floors 88 (bottom) and 89 (top): tuned mass damper observation floors.

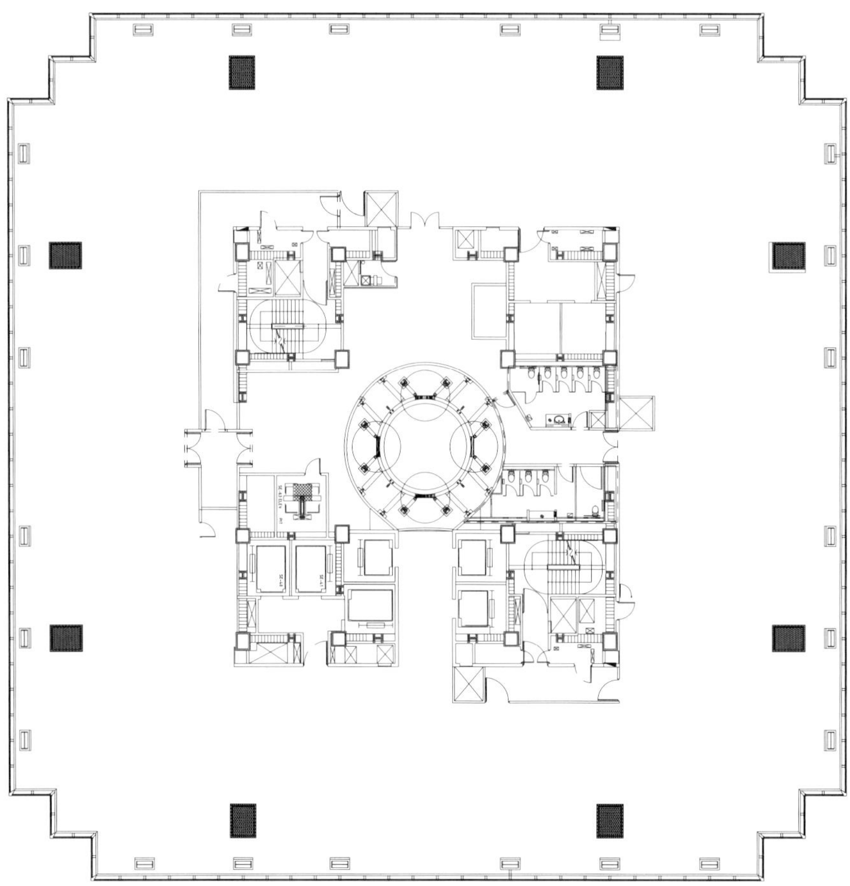

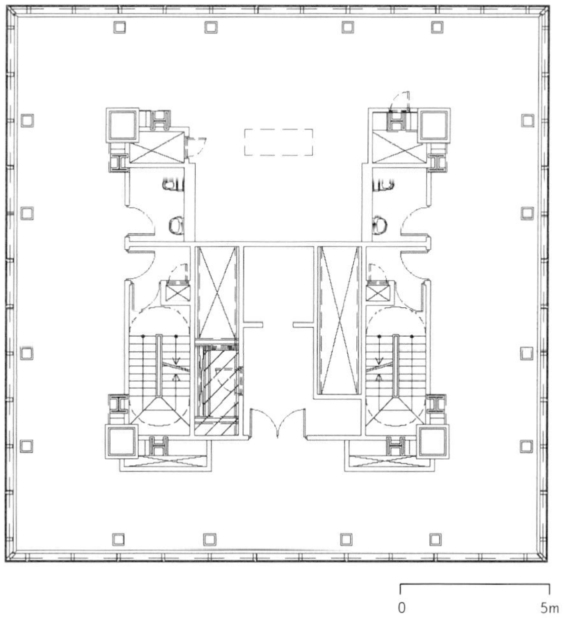

Floor 101: summit.

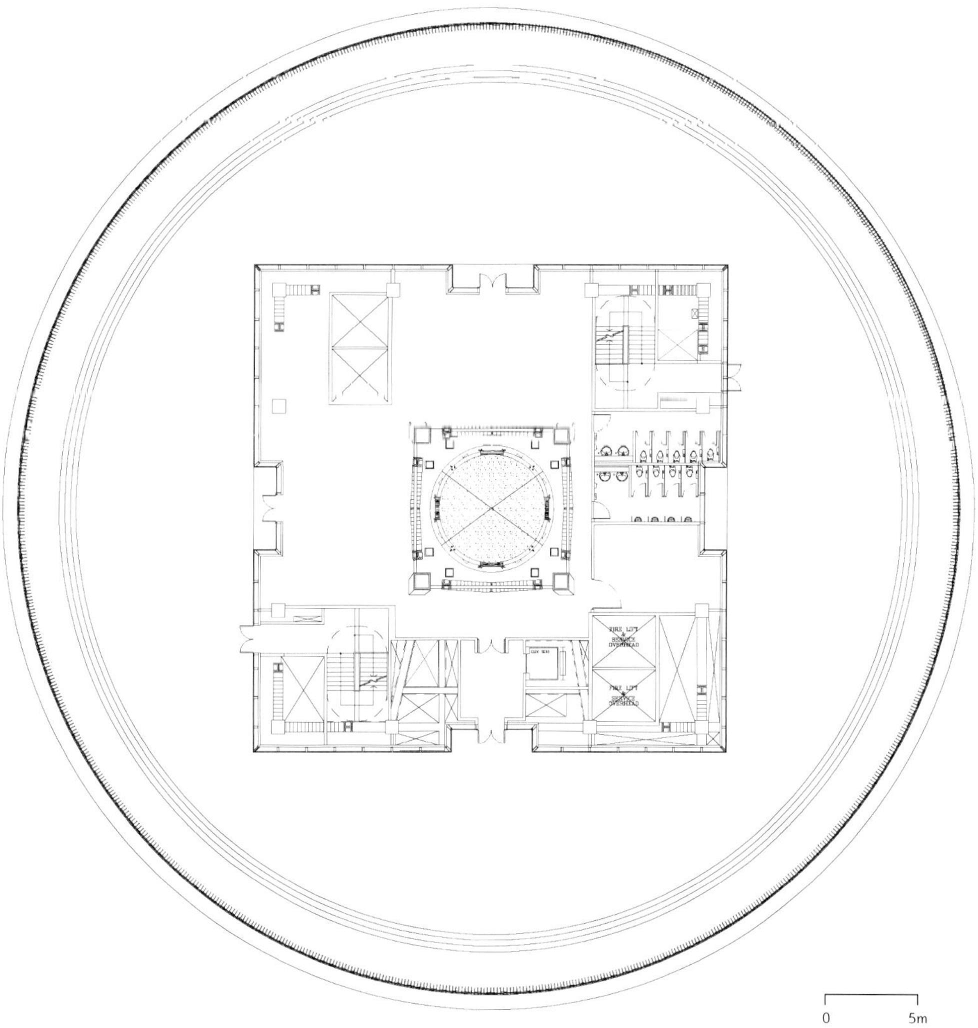

Floor 91: outdoor observation deck.

TECHNICAL DATA

Location	Taipei, Taiwan
District	Xin Yi Planned District
Street address	ShiFu Road / Xin Yi Road / SongZhi Road
Principal structural material	Steel
Other material	Glass curtain wall, stone wall base
Number of levels	101
Number of basements	5
Height	508 metres (1667 feet), structural top 448 metres (1470 feet), top of roof 438 metres (1437 feet), top occupied floor The tower stands on an elevated plaza 1 metre (3 feet) above the city sidewalks.
Site area	30,227 square metres/9159 pings
Tower foot print	4900 square metres
Total gross floor area	357,719 square metres/114,659 pings
Tower floor area	193,400 square metres/58,500 pings
59-84F High zone	Typical floor plate 2645-3570 square metres/800-1080 pings
35-58F Mid zone	Typical floor plate 2500-3305 square metres/760-1000 pings
9-34F Low zone	Typical floor plate /2810-4300 square metres/850-1300 pings
Mall floor area	77,033 square metres/23,000 pings
Car parking spaces	1839
Motorcycle parking spaces	2990
Elevators	61
Escalators	50

Office tower, office floors

Permitted use	Any business, including finance and insurance
Slab to slab height	4.20 metres
Internal clear height	2.80 metres
Curtain wall to core wall	9.80-13.89 metres
Curtain wall	Low-e, double glazed, aluminium framed
Mullion spacing	1.5 metres
Floor loading	Standard: 400 kilogram/square metre; high load areas: 1000 kilogram/square metre
Raised floor	6 centimetres high
Lighting level	650 lux average at desk level on open plan layout
Corners	Column-free corner views
Handover condition	Includes suspended ceiling, lighting, air conditioning and fire equipment

Chillers.

Vertical transportation

Passenger elevators:
10 double-deck elevators, 2040 kg (31 persons) per deck, shuttle elevators serving the transfer floors at 480 metres per minute
24 double-deck elevators, 1350 kg (20 persons) per deck, for access within 6 sub-zones (4 in each sub-zone) at 210 metres per minute
3 single-deck elevators (various capacities)
2 communication floors elevators
Observation elevators:
2 pressurised single-deck elevators, 1350 kg (20 persons) serving the observation decks at 1010 metres per minute and running a distance of 382.2 metres in 37 seconds
Service elevators:
3 single-deck elevators, 2 x 2040 kg, 1 x 4800 kg
Car park elevators:
6 single-deck elevators, 1600 kg to tower lobby

Mechanical and electrical

Fire control systems:
Integrated detection, extinguishing and pressurised evacuation systems designed to exceed all current standards.
Electrical supply:
Dual feed from two sub-stations, two risers for internal distribution, transformer redundancy. General office use design load of 65 VA/sq m (excluding air conditioning and lighting).
Power supply system:
Within the tower, separate dual risers distribute power to the nine mechanical floors and each office floor is connected to two mechanical floors, all of which minimises the risk of connection failure.
Emergency generators provide about 70 percent of the tower tenants' power consumption needs. An Uninterruptible Power Supply (UPS) ensures that the building's control and security systems remain in operation before the emergency generators start.

VAC system:
The VAC system consists of the chiller, the glycol chiller and the ice storage tank. The chiller produces chilled water for nighttime air conditioning. The glycol chiller pre-cools the glycol solution, which passes through the ice storage tank, to supply daytime air conditioning.
Variable Air Volume (VAV) boxes per floor.
Individual control provides ability to vary temperature and wind speed.
Separate 24-hour chilled water supply for computer rooms provided upon request.
VAC supply:
To ensure the optimum working environment, the design of the VAV system is based on world-class standards, the local weather patterns and future operating strategies.
Fire protection system:
The automatic sprinkler system is linked to the water tanks located in the mechanical plant rooms and is fed by gravity so that power failure cannot interrupt the water supply.
Each office floor has two pressurised fire safety corridors directly connected to separate pressurised emergency staircases, which provide access to the two refuge rooms located on each mechanical floor (every eight levels). Mechanical floors above the 34th floor have an outdoor refuge balcony. All the office floors and key escape routes are protected by smoke extractors.

VAKS kiosks in lobby.

Building management and control room.

Telecommunications

Chunghwa Telecom has installed the most advanced integrated telecommunications infrastructure. Two fibre-optic backbones, from separate Chunghwa Telecom central stations, enter the building at different points. Internal distribution from the Main Distribution Frames (MDFs) located on the 17th and 66th floors is via separate 10 Gbps backbones to an Intermediate Distribution Frame (IDF) located on each office floor.

TAIPEI 101 is equipped with three alternative means of communication: microwave, free space optics and satellite (to two earth-stations). The entire building is mobile accessible, including elevators and basements. The public areas are a wireless environment.

Broadband: 10 Gbps broadband from two access points and two risers for internal distribution

Mobile communication: connectivity throughout the building

Satellite communication: direct transmission to two satellite earth-stations

Microwave communication: stationed on level 34 to provide extra redundancy. Direct transmission to Chunghwa Telecom's central station.

Free space optics: stationed on level 18 to provide extra redundancy. Direct transmission to Chunghwa Telecom's central station.

Building management

Energy management and control system:

Developed by Siemens, the Energy Management and Control System (EMCS) tests and manages all energy related systems, such as air conditioning, lighting control, fire and life and safety systems. In addition to occupier safety, minimising the impact on the environment is a primary objective.

Security systems

Developed by Siemens, the security system - the central monitoring security center located on level B1 - controls access to the tower and monitors the safety of all equipment and occupants, using multiple tools including 420 CCTV cameras and smartcard access at all points of ingress and circulation.

Visitor Access Kiosk System (VAKS):

The Visitor Access Kiosks issue temporary cards for entering the tower to authorised visitors and keep photographic records.

Project team

Land owner	Taipei City Government
Developer	Taipei Financial Center Corp.
Shareholders	China Development Industrial Bank China United Trust & Investment Corp. Chunghwa Telecom Co., Ltd. Cathay Life Insurance Co., Ltd. Taiwan Stock Exchange Walsin Lihwa Corp. Mega Holdings Co., Ltd. Chie-Ho Engineering & Development Ltd. Shin Kong Life Insurance Co., Ltd. Chinatrust Commercial Bank Chinatrust Financial Holding Co., Ltd. China Life Insurance Co., Ltd First Financial Holding Co., Ltd. Da Fuo Shan Co., Ltd. Mega International Commercial Bank Hua Nan Financial Holdings Co., Ltd.
Architect	C.Y. Lee & Partners Architects/Planners
Structural engineer	Evergreen Consulting Engineering
Structural engineer (peer review role)	Thornton Tomasetti
Wind damper consultant	Motioneering Inc.
Wind tunnel	RWDI
Curtain wall	ALT Design
Soil test engineer	Sino Geothechnology, Inc.
MEP engineer	Continental Engineering Consultants, Inc.
MEP engineer (Peer review role)	Lehr Associates
Vertical transportation consultant	Lerch, Bates & Associates Inc.
Acoustic engineer	SMW
Landscape architect	Genius Loci
Telecommunications system	Chunghwa Telecom
Building management system	Siemens
Lighting	Theo Kondos
Project manager	Turner International
Contractor	KTRT (Kumagai Gumi; Taiwan Kumagai; RSEA Engineering; Ta-You-Wei Construction)
Steel structure	China Steel, Nippon Steel
Elevators and escalators	Toshiba Elevator and Building Systems Corp. installed by GFC, Ltd.

Honours and records

- Tallest structural top at 508 metres confirmed in 2004 by CTBUH (Council on Tall Buildings and Urban Habitat, Chicago, Illinois, USA)
- Tallest roof at 448 metres confirmed in 2004 by CTBUH
- Tallest occupied floor at 438 metres confirmed in 2004 by CTBUH
- World's fastest elevator confirmed by Guinness World Records (observatory elevator, 1010 metres per minute or 60.6 kilometres per hour)
- The world's largest exposed wind damper, weighing 660 metric tons
- Named top of '7 Wonders of Engineering' by Discovery Channel
- Named 'Best of What's New 2004' by *Popular Science* magazine

From here, all you see was sculpted, piece by piece,
through years of devotion.
Dedicated to all the partners in TAIPEI 101 Project.

At this moment still many partners
Following the tracks of the forerunners
Continue the story of TAIPEI 101
A never-ending story.

– inscription from Partners' monument

Photography and drawing credits

Unless otherwise indicated, all photographs and drawings are copyright of Taipei Financial Center Corp.

Page 9: courtesy The Skyscraper Museum, New York, New York

Page 10: Coll. G Binder/Buildings & Data SA

Page 11: Frankie FY Wong

Page 12: Frank P Palmer

Page 13 rendering: MPai+, Hong Kong

Page 14 (l): rendering: dbox, courtesy Silverstein Properties

Page 14 (r): courtesy C.Y. Lee & Partners Architects/Planners

Page 18 (b): courtesy C.Y. Lee & Partners Architects/Planners

Page 21: courtesy C.Y. Lee & Partners Architects/Planners

Page 26: courtesy C.Y. Lee & Partners Architects/Planners

Page 28: courtesy C.Y. Lee & Partners Architects/Planners

Page 30: courtesy C.Y. Lee & Partners Architects/Planners

Page 31: courtesy C.Y. Lee & Partners Architects/Planners

Page 35 (inset): ©Chunni4691, Dreamstime.com

Page 36 (bamboo): ©Alexander Tivanov, Dreamstime.com

Page 37 (r): ©Kim Reinick, Dreamstime.com

Page 47 (t): © Kentoh, Dreamstime.com

Pages 168-177: courtesy C.Y. Lee & Partners Architects/Planners

Taipei Financial Center Corp.

59F TAIPEI 101 TOWER,

7 Xin Yi Road Sec 5, Taipei 110, Taiwan

TEL: + 886 2 8101 7777

FAX: + 886 2 8101 8949

www.taipei-101.com.tw